Fluid Mechanics and Thermo-Acoustic Waves

Fluid Mechanics and Thermo-Acoustic Waves

Timothy S. Margulies

ISBN: Softcover 978-1-4363-9901-2

This book was printed in the United States of America.

To order additional copies of this book, contact:
Xlibris Corporation
1-888-795-4274
www.Xlibris.com
Orders@Xlibris.com

57401

CONTENTS

Dedicated to

Researchers who endeavor to build knowledge and understanding through a unified framework of mathematics and scientific experiments.

Students and professionals of all ages who wish to be self-taught while striving to improve their daily lives, products, and services or who just have a love of learning.

The family, in particular, my mother and father.

Acknowledgement:

The author gratefully acknowledges the interactions with Professor Abdul Siddiqui (Penn State York) who suggested the relook at the wire coating and simple fluid stress tensor problem. Further the encouragement and collaborative help of A. Benharbit (Penn State York) is noteworthy on acoustic waves through particulate suspensions. The manuscript has genuinely benefited as well from continued discussions with Professor Robert Powell (University of California at Davis) through the this undertaking.

I

Generalized Hagen-Poiselle Motions: Motion of an Incompressible Visco-Elastic through a Tube Using a Fractional Calculus Rheology Model

Here is presented a derivation for the steady state motion of a visco-elastic through a horizontal tube of radius. R. Cauchy's first law for an incompressible may be written as $\rho\left[\frac{\partial^2 \underline{\underline{\mathbf{u}}}}{\partial t^2}+\left(\nabla\frac{\partial \underline{\underline{\mathbf{u}}}}{\partial t}\right)\bullet\frac{\partial \underline{\underline{\mathbf{u}}}}{\partial t}\right]=-\nabla P+\nabla\bullet\underline{\underline{\mathbf{T}}}$, where the modified pressure $P \equiv p+\rho\phi$ assumes a uniform potential ϕ such as for an external gravitational force ($-\nabla\phi$) with $\varphi = gz$, and $\underline{\underline{S}} = -p\underline{\underline{I}}+\underline{\underline{T}}$.

$$0 = -\frac{\partial P}{\partial r}$$

$$0 = -\frac{1}{r}\frac{\partial P}{\partial \theta}$$

$$\frac{\partial P}{\partial z} = \frac{1}{r}\frac{d}{dr}(rT_{rz})$$

At $z = 0$, $r = R$, $\theta = 0$, $-S_{rr} = P_0$ and at $z = L$, $r = R$, $\theta = 0$, $-S_{rr} = P_L$ so $\Delta P = P_L - P_0$. The pressure drop between positions along the axis z of the tube separated by length L apart is related to the radial location r by the formula, $-\frac{\Delta P}{L}\frac{r}{2} = T_{rz}$, where T_{rz} is the rz shear component of the extra stress tensor (i.e., in the direction of z perpendicular to r). Let the material stress response

$T_{rz} = E_0 \frac{\partial u_z}{\partial r} + E_1 \frac{\partial^\alpha \partial u_z}{\partial t^\alpha \partial r}$ with u_z the axial displacement for deformation of the body, and $\frac{\partial u_z}{\partial r}$ denoting the strain. A similar generalized Voigt model introduced by Caputo using fractional derivatives, $T(t) = E_0\varepsilon(t) + E_1 \frac{d^\alpha}{dt^\alpha}\varepsilon(t)$; $E_0, E_1 \geq 0$ also possesses a frequency-independent specific dissipation function with possible association to the phenomenon of fatigue. Here, T is the time-dependent extra stress, and ε denotes the strain, which also depends on time. The fractional derivative α has been found useful in many applications:

$$\frac{\partial^\alpha f}{\partial t^\alpha} \equiv \frac{1}{\Gamma(1-\alpha)} \frac{d}{dt} \int_0^t \frac{f(\tau)d\tau}{(t-\tau)^\alpha} \quad 0 < \alpha < 1$$

The motion is assumed to be laminar and parallel to the tube walls with $\phi = 0$. At the tube wall, r = R, the motion is zero. Integrating from the boundary radius R to the interior radius r obtains,

$$-\frac{\Delta P}{2L}\int_R^r r \cdot dr = \int_0^{u_z^*} \left\{ E_0 + E_1 \frac{\partial^\alpha}{\partial t^\alpha} \right\} \partial u_z$$

or $\frac{-\Delta P}{4L}\left(r^2 - R^2\right) = E_0 u_z^* + E_1 \frac{\partial^\alpha u_z^*}{\partial t^\alpha}$. Rewriting using

$a \equiv \frac{E_0}{E_1}$, $b(r) \equiv \frac{-\Delta P}{4LE_1}\left(r^2 - R^2\right)$ yields $\frac{\partial^\alpha u_z^*}{\partial t^\alpha} + a \cdot u_z^* = b(r)$.

For the special case of a = 0 and α = 0 with $E_1 = \mu$, the shear viscosity, the Hagen-Poiselle fluid model is developed for a Newtonian fluid. Next, examine the case for α = 0.5 (i.e., sometimes referred to as semi-differentials). Substituting this value results in a differential equation defining the problem for this material, which can be solved assuming a boundary condition such as no slip at the wall, $u_{z=0}$ at r = R.

The homogeneous solution for a constant C is given by,

$$u_z^* = C \cdot L^{-1}\left[\frac{1}{s^{0.5} + a}\right] = C \cdot \left\{ \frac{1}{\sqrt{\pi t}} - a \cdot \exp\left(a^2 t\right) \cdot \mathrm{erfc}\left(a\sqrt{t}\right) \right\}$$

and the particular solution by $\frac{b}{a}=\frac{-\Delta P}{4LE_0}(r^2-R^2)$. Solve for C by using the stated no-slip boundary condition. Combining the solutions forms,

$$u_z^* = \cdot\left\{\frac{1}{\sqrt{\pi t}} - a \times \exp\left(a^2 t\right) \times \operatorname{erfc}\left(a\sqrt{t}\right)\right\}\left\{1-\sqrt{\pi t}\cdot a \cdot \exp\left(a^2 t\right)\cdot \operatorname{erfc}\left(a\sqrt{t}\right)\right\} + \frac{-\Delta P}{4LE_0}\left(r^2 - R^2\right)$$

Therefore, it is recommended that experiments be performed to test the solid-fluid visco-elastic motion represented by fractional calculus to explore and to validate its usefulness.

To further investigate this matter, consider again, a fractional calculus material representation applied to polymer solutions, $T_{rz} = E_0 + E_1 \frac{\partial^{\alpha+1} \partial u_z}{\partial t^{\alpha+1} \partial r}$. The previous differential equation becomes for this type of generalized Bingham stress model,

$\frac{\partial^{\alpha+1} u_z^*}{\partial t^\alpha \partial r} = -\frac{\Delta Pr}{2LE_1} - \frac{S_0}{E_1}$ where upon integrating with respect to r,

$\frac{\partial^\alpha u_z^*}{\partial t^\alpha} = \left[-\frac{\Delta Pr^2}{4LE_1} - \frac{S_0 r}{E_1}\right]_R^r = -\frac{\Delta P}{4LE_1}\left(r^2 - R^2\right) + \frac{S_0}{E_1}(R-r)$. The homogeneous solution is found from the inverse Laplace transform $L^{-1}\left\{\frac{1}{s^\alpha}\right\}$, or $u_z^* = \frac{t^{\alpha-1}}{\Gamma(\alpha)}$. For illustration, assume that $\alpha = 0.5$ and form a particular solution of the polynomial form $u_z^* = At^{0.5}$ for constant A. Substituting into the differential equation with the semi-differentiation allows evaluation of the constant with the nonhomogeneous terms.

$$A = -K_{1/2}^{-1}\left(-\frac{\Delta P}{4LE_1}\left(r^2 - R^2\right) + \frac{S_0}{E_1}(R-r)\right), \text{ since}$$

$$\frac{d^{0.5}}{dx^{0.5}} \cdot x^{n+\frac{1}{2}} = K_n \cdot x^n, \quad K_n \equiv \frac{(2n+1)\cdot!\sqrt{\pi}}{2\cdot 4^n \cdot [n!]^2}.$$

The general solution is the sum of the complementary and particular solutions.

Let the volume flow rate be designated $Q = \int_0^{2\pi}\int_0^R \left(\frac{\partial u_z}{\partial t}\right) r \cdot dr \cdot d\theta$ which equals, $-2\pi\int_0^R \left(\frac{\partial u_z}{\partial t}\right) r \cdot dr$. Therefore, integrating by parts obtains, $Q = -\pi\int_0^R \frac{d}{dr}\left(\frac{\partial u_z}{\partial t}\right) r^2 dr$. Defining new variables, $\tau = \frac{r}{R}\tau_R = \frac{r}{R}\left(-S_{rz}\big|_{r=R}\right)$, rewrites $Q = -\pi\left(\frac{R^3}{\tau_R^3}\right)\int_0^{\tau_R} \frac{d}{dr}\left(\frac{\partial u_z}{\partial t}\right)\tau^2 d\tau$. Differentiating both sides, $\frac{d\left(Q\tau_R^3\right)}{d\tau_R} - \pi R^3 \tau_R^2 \frac{d}{dr}\left(\frac{\partial u_z}{\partial t}\bigg|_{r=R}\right)$. This provides formulas to explore materials by measuring pressure drop versus volume of flow and analyzing for the stress at the wall τ_R versus strain rate.

References

Slattery, John C. *Momentum, Energy, and Mass Transfer in Continua.* Florida: R. Krieger Publishing, 1981.

Tanner, Roger I. *Engineering Rheology.* New York: Oxford University Press, 1985.

Caputo, M. *Geophysical Journal of Royal Astronomical Society* 18 (1967): 529-39.

Caputo, M. *Journal of the Acoustical Society of America* 66 (1979): 176-9.

Oldham, Keith B. and Jerome Spanier. *The Fractional Calculus.* : Dover, 2006.

O'Neil, Peter V. *Advanced Engineering Mathematics.* California: Wadsworth Inc., 1983.

S. P. Sutera and R. Skalak, *in Annual Rev. Fluid Mech,* 25: 1-19, 1993.

II

Rayleigh's Problem of Oscillating Plate Boundary Motions

Consider an infinite, incompressible visco-elastic overlying a wall using coordinates (z_1, z_2, z_3) with gravity in the negative z_2 direction. The wall or plate is bounded on the one side by the (z_1, z_3) plane. At time $t > 0$ the wall is set into motion in the z_1 direction with a sinusoidal perturbation as follows:

$u_1 = A\cos(\omega t - \varepsilon)$. For all t, $z_2 \to \infty$, $\frac{\partial p}{\partial z_1} = \frac{\partial p}{\partial z_2} = 0$. At $t = 0$, $\underline{v} = 0 \ \forall z_2$ and

$v_1 = v_1(z_2, t)$. $\rho \frac{\partial^2 u_1}{\partial t^2} = \frac{\partial}{\partial z_2}\left[E_0 \frac{\partial u_1}{\partial z_2} + E_1 \frac{\partial^\alpha \partial u_1}{\partial t^\alpha \partial z_2}\right] = E_0 \frac{\partial^2 u_1}{\partial z_2^2} + E_1 \frac{\partial^\alpha \partial^2 u_1}{\partial t^\alpha \partial z_2^2}$.

Let $u_1 = u(z_2)\mathrm{ep}(i(\omega t - \varepsilon))$ then

$-\rho\omega^2 u = E_0 \frac{\partial^2 u}{\partial z_2^2} + E_1 (i\omega)^\alpha \frac{\partial^2 u}{\partial z_2^2} = \left(E_0 + E_1 (i\omega)^\alpha\right)\frac{\partial^2 u}{\partial z_2^2}$. Rewriting,

$u'' + \left(\frac{\omega^2}{\frac{E_0}{\rho} + \frac{E_1 (i\omega)^\alpha}{\rho}}\right) u = 0$. This second order ordinary differential equation can be solved. Remembering in general that the equation $y'' + ay' + by = 0$ using dependent variable y as a function of x has complex roots m_1, m_2 if $a^2 - 4b < 0$

$$\alpha = \frac{-a}{2}, \qquad \beta = \frac{1}{2}\sqrt{4b - a^2} = \sqrt{b} = \left(\frac{\omega^2}{\frac{E_0}{\rho} + \frac{E_1 (i\omega)^\alpha}{\rho}} \right)^{0.5};$$

$$m_1 = \alpha + i\beta, \qquad m_2 = \alpha - i\beta\ ,$$

where $y(x) = e^{\alpha x}(C_1 \cos(\beta x) + C_2 \sin(\beta x))$. Here, in the above application, $a = 0,\ \alpha = 0;\ m_1 = ik,\ m_2 = -ik.$

Another generalized stress model using a fractional calculus expression is incorporated into the equation of motion follows,

$$\rho \frac{\partial^2 u_1}{\partial t^2} = \frac{\partial}{\partial z_2}\left[E_0 u_1 + E_1 \frac{\partial^\alpha \partial u_1}{\partial t^\alpha \partial z_2} \right] = E_0 \frac{\partial u_1}{\partial z_2} + E_1 \frac{\partial^\alpha \partial^2 u_1}{\partial t^\alpha \partial z_2^2}.$$

Again let the solution be of the form $u_1 = u(z_2)\exp(i(\omega t - \varepsilon))$ which obtains,

$-\rho\omega^2 u = E_0 \frac{\partial u}{\partial z_2} + E_1 (i\omega)^\alpha \frac{\partial^2 u}{\partial z_2^2}$. Then $\frac{\partial^2 u}{\partial z_2^2} + \frac{E_0}{E_1 (i\omega)^\alpha}\frac{\partial u}{\partial z_2} = \frac{-\rho\omega^2}{E_1 (i\omega)^\alpha} u$. Let

$$\lambda^2 \equiv \frac{\rho\omega^2}{E_1 (i\omega)^\alpha},$$

$\frac{\partial^2 u}{\partial z_2^2} + \frac{E_0}{E_1 (i\omega)^\alpha}\frac{\partial u}{\partial z_2} + \lambda^2 u = 0$. Solving the auxiliary equation

$m^2 + am + b = 0,\ a \equiv \frac{E_0}{E_1 (i\omega)^\alpha} \quad b \equiv \lambda^2$ yields two roots

$$m_1 = p + iq,\ m_2 = p - iq \neq m_1$$

Then $u = e^{pz_2}(C_1 \cos(qz_2) + C_2 \sin(qz_2))$ or defining

$$\mu = \sqrt{(C_1)^2 + (C_2)^2},\ \tan\delta = C_1 / C_2,\ u = \mu e^{pz_2} \sin(qz_2 + \delta)$$

At $z_2 = 0,\ u_1 = u_0 \cdot \exp(i(\omega t - \varepsilon)),\ u(z_2 = 0) \equiv u_0$ [known] and the velocity remains finite as $z_2 \to \infty$.

III

Visco-Elastic Lubrication

Consider planar motion with horizontal and vertical displacements given by u and w, respectively. The balance of mass equation for an incompressible becomes

$$\frac{\partial u}{\partial x}+\frac{\partial w}{\partial z}=0 \ .$$

The equations of motion neglecting body forces are written,

$$\frac{-\partial p}{\partial x}+\frac{\partial T_{xx}}{\partial x}+\frac{\partial T_{xz}}{\partial z}=\rho\cdot\ddot{u}$$

$$\frac{-\partial p}{\partial z}+\frac{\partial T_{xz}}{\partial x}+\frac{\partial T_{zz}}{\partial z}=\rho\cdot\ddot{w}$$

For small gaps between surfaces compared to length scales, O (h/L) approximates the equations of motion as follows:

$$\frac{-\partial p}{\partial x}+\frac{\partial T_{xz}}{\partial z}=0$$

$$\frac{-\partial p}{\partial z}=0$$

The second equation above shows that $p = p\,(\,x\,)$ while the first equation may be integrated with respect to z,

$$T_{xz}=-\frac{\partial p}{\partial x}z+C(x)$$

The Newtonian viscous model, $T_{xz} = \eta_0 \frac{\partial \dot{u}}{\partial z}$, may be generalized by a fractional calculus one, such as $T_{xz} = E_0 \frac{\partial u}{\partial z} + E_1 \frac{\partial}{\partial z}\frac{\partial^\alpha u}{\partial t^\alpha}$. Substituting into the momentum equation gives,

$$E_0 \frac{\partial u}{\partial z} + E_1 \frac{\partial}{\partial z}\frac{\partial^\alpha u}{\partial t^\alpha} = -\frac{\partial p}{\partial x} z + C(x)$$

integrating, $E_0 \cdot u + E_1 \frac{\partial^\alpha u}{\partial t^\alpha} = -\frac{\partial p}{\partial x} \cdot \frac{z^2}{2} + C(x) \cdot z + D$

The boundary conditions, $u(z=0) = U_0$, $u(z=h) = U_1$, are applied to the case of $\alpha = 0.5$ for illustration with results

$$D = E_0 U_0 + E_1 \frac{\partial^{\alpha=0.5} U_0}{\partial t^{0.5}} = E_0 U_0 + E_1 \frac{U_0}{\sqrt{\pi t}} = U_0 \left(E_0 + E_1 \frac{1}{\sqrt{\pi t}} \right)$$

$$C = -\frac{h}{2}\frac{\partial p}{\partial z} + \frac{U_1 (U_1 - U_0)}{h} \left(E_0 + E_1 \frac{1}{\sqrt{\pi t}} \right)$$

The homogeneous solution is similar to the earlier equation investigated.

IV

Helical Flow

The equations of motion in cylindrical coordinates decouple. Assume a constant pressure gradient, $-\Pi \equiv T_{zz,z}$ then, $\frac{\partial T_{rz}}{\partial r} + \frac{1}{r} T_{zr} = -\Pi$. Multiplying by r and rewriting, $\frac{\partial}{\partial r}(r \cdot T_{zr}) = -\Pi \cdot r$ Integrating both sides of the equation, gives $r T_{zr} = -\Pi \frac{r^2}{2} + a(cst)$. Therefore, $T_{zr} = -\Pi \cdot \frac{r}{2} + \frac{A}{r}$.

Therefore, a no-slip or zero shear stress at the boundary $r = r_1$ and $r = r_2$ obtains $A = \dfrac{\Pi (r_2 - r_1)}{2 \ln\left(\dfrac{r_2}{r_1}\right)}$.

Let the fractional calculus extra stress rheology be represented by $T_{zr} = E_0 \frac{\partial u_z}{\partial r} + E_1 \frac{\partial}{\partial r} \frac{\partial^\alpha u_z}{\partial t^\alpha}$. Combining the last two equations,

$$E_0 \frac{\partial u_z}{\partial r} + E_1 \frac{\partial}{\partial r} \frac{\partial^\alpha u_z}{\partial t^\alpha} = -\Pi \cdot \frac{r}{2} + \frac{A}{r}, \qquad \frac{\partial}{\partial r}\left(E_0 u_z + E_1 \frac{\partial^\alpha u_z}{\partial t^\alpha} \right) = -\Pi \cdot \frac{r}{2} + \frac{A}{r} .$$

Integrating, $E_0 u_z + E_1 \frac{\partial^\alpha u_z}{\partial t^\alpha} = -\Pi \cdot \frac{r^2}{4} + A \ln r \Big|_{r_1}^{r}$

Examine the example $\alpha = 0.5$ case where the homogeneous equation becomes a previously examined equation, $\frac{\partial^\alpha u_z}{\partial t^\alpha} + \frac{E_0}{E_1} u_z = 0$ The complementary solution becomes,

$$u_z^* = C \cdot L^{-1}\left[\frac{1}{s^{0.5} + a}\right] = C \cdot \left\{\frac{1}{\sqrt{\pi t}} - a \cdot \exp\left(a^2 t\right) \cdot \operatorname{erfc}\left(a\sqrt{t}\right)\right\}, \quad a = \frac{E_0}{E_1}$$

Knowing

$$\frac{d^{0.5}}{dx^{0.5}} \cdot x^{n+0.5} = K_n x^n, \; K_n = \frac{(2n+1)!\sqrt{\pi}}{2 \cdot 4^n [n!]^2}, \; \frac{d^{0.5}}{dx^{0.5}} \ln(x) = \frac{\ln(4x)}{\sqrt{\pi x}}, \; \frac{d^{0.5}}{dx^{0.5}} cst = \frac{cst}{\sqrt{\pi x}}$$

the particular solution can be found. The volume flow rate is

$$Q = \int_0^{2\pi} \int_{r_1}^{r} \dot{u}_z(r) r\, dr\, d\theta .$$

Now turning to investigating the circumferential (or rotary) θ momentum equation,

$$\frac{2}{r} T_{\theta r,\theta} + \frac{\partial}{\partial r} T_{\theta r,r} = 0, \quad \frac{\partial}{\partial r}\left(r^2 T_{\theta r}\right) = 0, \quad \Rightarrow \left(r^2 T_{\theta r}\right) = B(cst).$$

$T_{r\theta} = \frac{B}{r^2}$. Substituting for the stress, $T_{r\theta} = r \frac{\partial}{\partial r}\left(E_0\left(\frac{u_\theta}{r}\right) + E_1 \frac{\partial^\alpha\left(\frac{u_\theta}{r}\right)}{\partial t^\alpha}\right)$,

and then integrating with respect to r,

$$\int_{R_1}^{r} \frac{B}{r^3} dr = \frac{-B}{2}\left(\frac{1}{r^2} - \frac{1}{R_1^2}\right) = \int_{R_1}^{r} d\left(E_0\left(\frac{u_\theta}{r}\right) + E_1 \frac{\partial^\alpha\left(\frac{u_\theta}{r}\right)}{\partial t^\alpha}\right)$$

which can be solved with the rotary boundary conditions.

V

Acoustic Horns Exponential, Power Law, Gaussian Shapes

Abstract

The purpose of this letter is to present a theory of a generalized "Webster" horn equation for a diverse class of visco-elastic materials represented by fractional calculus rheology models, which have previously not been investigated. The acoustic wave propagation pertains to a one-dimensional infinite, homogeneous horn, duct, or rod with small area change along the longitudinal x-axis. Solution techniques are given for area functions in the form of a power law, exponential, and Gaussian cross-sectional area shapes. Both the shape and material properties may provide design options for amplification or for attenuation and damping depending on the application.

First, using the area averaged density and stress, the relevant continuum balance equations are given. They are specialized for this investigation for a fractional calculus rheology stress model. These models have offered simple efficient representations for the stress response of many polymers. Seeking a solution separating periodic time and space functions obtains a generalized "Webster" horn equation for this more general class of viscoelastic materials. The theory is then applied to area functions in the form of a power law, an exponential, and a Gaussian cross-sectional shape. Defining the averaged density $\bar{\rho}$ and stress $\bar{S}$ for an area function A(x) along the longitudinal x-axis:

$$\bar{\rho} = \rho A(x)$$
$$\bar{S} = A(x)S$$

The balance of total mass with no supply terms may be written,

$$\frac{D\bar{\rho}}{Dt} = \frac{\partial\bar{\rho}}{\partial t} = \frac{\partial\rho}{\partial t} = 0$$

The balance for linear momentum for this area average system obtains for stress tensor $\bar{S}_x$ and displacement u whose time differential is the velocity v (in the x-direction) :

$$\bar{\rho}\frac{Dv}{Dt} = \bar{\rho}\frac{\partial v}{\partial t} = \bar{\rho}\frac{\partial^2 u}{\partial t^2} = \frac{\partial}{\partial x}\bar{S}_{xx} = \bar{S}_{xx,x}$$
$$\rho A\frac{\partial^2 u}{\partial t^2} = (AS_{xx})_{,x} = A_{,x}S_{xx} + AS_{xx,x} \quad \text{velocity } v \equiv \frac{\partial u}{\partial t}$$

The equation of motion above obtains the simple form:

$$\rho\frac{\partial^2 u}{\partial t^2} = \frac{A_{,x}}{A}S_{xx} + S_{xx,x}$$

The stress response of the material may be represented in terms of strain $\frac{\partial u}{\partial x}$ for an elastic, or of its time differential, rate of strain for a visco-elastic. The following may be analyzed,

$$S_{xx} = E_0\frac{\partial u}{\partial x} \qquad \text{Elastic}$$
$$S_{xx} = E_0\frac{\partial u}{\partial x} + E_1\frac{\partial^{\alpha+1}u}{\partial t^{\alpha}\partial x} \qquad \text{Viscoelastic}$$

for $\alpha = 1$, $S_{xx} = E_0\frac{\partial u}{\partial x} + E_1\frac{\partial^2 u}{\partial t\partial x}$

Here, the fractional derivative $0<\alpha<1$ is defined as,

$$\frac{\partial^\alpha f}{\partial t^\alpha} \equiv \frac{1}{\Gamma(1-\alpha)} \frac{d}{dt} \int_0^x \frac{f(\tau)d\tau}{(t-\tau)^\alpha}]$$

$$\Gamma(x) \equiv \int_0^\infty t^{x-1} e^t dt \quad (\text{Re } x > 0)$$

$$\Gamma(x) \equiv -\frac{1}{2i\sin\pi x} \int_c (-t)^{x-1} e^t dt \quad x \text{ non-integer}$$

This fractional calculus stress for elastico-viscous response applies to many polymer solutions with $\alpha=0.5$ (i.e., sometimes referred to as semi-differentials) thereby simplifying the previous expressions with a larger number of terms for material stress response representation in several cases examined. From a Fourier transform approach has given $\frac{\partial^\alpha}{\partial t^\alpha}(e^{i\omega t}) = (i\omega)^\alpha (e^{i\omega t})$. More generally,

$$\frac{\partial^\alpha}{\partial t^\alpha}(e^{i\omega t}) = (\omega)^\alpha \left(e^{i(\omega t + \frac{\pi\alpha}{2})} \right) \quad \text{for } \omega \geq 0.$$

Proceeding to solve the horn equation, let $u = G(x)e^{i\omega t}$

For the elastic case, $G'' + \frac{A_{,x}}{A} G' + k^2 G = 0 \quad c^2 = \frac{E_0}{\rho}, \quad k = \frac{\omega}{c}$

For phase speed c and wave number k.

The fractional derivative visco-elastic model yields,

$$G'' + \frac{A_{,x}}{A} G' + \frac{\rho\omega^2}{E_0 + E_1 (i\omega)^\alpha} G = 0$$

Consider the area shape function A(x) of the power-law form, $A_0(1+px)^n$ where $n = 1$ corresponds to a linear variation in cross-section area, and $n = 2$ models a conical case. Then

$$\frac{A'(x)}{A(x)} = \frac{pn}{(1+px)}$$

Transforming to variable y,

$$y=\left(\frac{1}{p}+x\right),\ \frac{\partial(\,)}{\partial y}\equiv(\,)^{\bullet},\ \kappa^2\equiv\frac{\omega^2}{\frac{E_0}{\rho}\left\{1+\frac{E_1}{E_0}(i\omega)^{\alpha}\right\}}$$

$$G^{\bullet\bullet}+\frac{n}{y}G^{\bullet}+\kappa^2 G=0$$

This equation has solutions in the form of, for example Hankel functions or Bessel's functions of the third kind.

Defining $n = 2\nu + 1$, or $\nu = \frac{n-1}{2}$,

$$G=\{\ \Pi_1 H_\nu^{(1)}(\kappa y)+\Pi_2 H_\nu^{(2)}(\kappa y)\ \}\ /\ y^{\nu}$$

$$H_\nu^{(1)}=J_\nu+iY_\nu\quad H_\nu^{(2)}=J_\nu-iY_\nu\quad \frac{d}{dx}H_\nu^{(1,2)}=-H_{\nu+1}^{(1,2)}$$

By examining asymptotic forms for these special functions and considering left-going waves, one sets $\Pi_2=0$. Now applying a boundary condition at x = L helps solve for the

constant Π_1. $P_0 e^{i\omega t}=S_{xx}A(x=L)\ =\ \left\{E_0\frac{\partial u}{x}+E_1\frac{\partial^{\alpha+1}u}{\partial t^{\alpha}\partial x}\right\}A_L$

Since $G'\big|_{x=L}=\frac{P_0}{E_0+E_1(i\omega)^{\alpha-1}}\quad \Pi_1=\frac{-P_0}{\kappa H^1_{\nu+1}\left(E_0+E_1(i\omega)^{\alpha-1}\right)}$

For an exponential shape function, $A(x)=A_0 e^{\frac{2x}{L}}$, the generalized horn equation becomes with $\frac{A'}{A}=\frac{2}{L}$, $G''+\frac{2}{L}G'+\kappa^2 G=0$.

For solutions of the form $G = e^{mx}$,

$$m^2+\frac{2}{L}m+\kappa^2=0,\ m_{1,2}=\frac{1}{L}\left\{-1\pm\sqrt{1-\kappa^2L^2}\right\}$$

$$G=a_1e^{m_1x}+a_2e^{m_2x}\qquad a_2=0$$

This solution corresponds to left-going waves. Solving for the constant a_1 with the boundary condition at L,

$$S_{xx}A(x=L)=P_0e^{i\omega t} \quad u=G(x)e^{i\omega t}$$

$$a_1=\frac{P_0}{\left(E_0+E_1(i\omega)^{\alpha-1}A_L m_1 e^{m_1L}\right)}$$

This solution captures an exponential horn shape function using a fractional calculus rheology model for stress reactions.

The dynamic equation of motion for a Gaussian horn in the x-direction is given by,

$$\Gamma''-\frac{2}{L^2}x\cdot\Gamma'+k^2\Gamma=0, \quad \text{let} \quad x\equiv Lz \quad \text{then} \quad \frac{d}{dx}=\frac{1}{L}\frac{d}{dz}, \ \frac{d^2}{dx^2}=\frac{1}{L^2}\frac{d^2}{dz^2}$$

$$\frac{d^2\Gamma}{dz^2}-2z\frac{d\Gamma}{dx}+2n\Gamma=0, \quad \text{where} \quad n\equiv 0.5k^2L^2$$

This transformed equation is a Hermite equation, which admits Hermite polynomial solutions. Assume a power series solution, form the derivatives needed to substitute into the equation to obtain a relation among the coefficients a_k and a_{k+2} by equating each coefficient term separately to zero:

$$\text{Let} \quad H(z)=\sum_{k=0}^{\infty}a_kz^k=a_0+a_1z+a_2z^2+a_3z^3+\ldots$$

$$H'(z)=\sum_{k=1}^{\infty}ka_kz^{k-1}=a_1+2a_2z+3a_3z^2+..$$

$$H''(z)=\sum_{k=2}^{\infty}k(k-1)a_kz^{-2k}=2a_2+3\cdot 2a_3z^2+4\cdot 3a_4z^2+..$$

A recursion relation can be developed $a_{k+2}=\dfrac{2n-2k}{(k+1)(k+2)}a_k$. Let a_0 and a_1 be two arbitrary constants, then

$$H(x)=a_0\left(1+\frac{a_2}{a_0}z^2+\frac{a_4}{a_2}\frac{a_2}{a_0}z^4+\ldots\right)+a_1\left(z+\frac{a_3}{a_1}z^3+\ldots\right).$$

This writing separates into even and odd powers (or parity). Typical tabulated values correspond to setting the coefficient of the largest term as 2^n. It is noted that

$$H_n(z) = (-1)^n e^{z^2} \frac{\partial^n}{\partial z^n} e^{-z^2}, \quad H'_n(z) = 2nH_{n-1}(z)$$

H(x) diverges as x->∞, $\frac{a_{k+2}}{a_k} \to \frac{2}{k}$. Thus a finite sum of terms is used. For n odd, set $a_0=0$; and for n even, set $a_1=0$.

The solution $\Gamma = KH(z)$ for constant A is applied to the boundary condition at x = L.

$$P_0 e^{i\omega t} = S_{xx} \cdot A = \frac{1}{L}\left\{E_0 \frac{\partial u}{\partial z} + E_1 \frac{\partial^{\alpha+1} u}{\partial t^{\alpha+1} \partial z}\right\} A_L \quad \text{where } u = \Gamma e^{i\omega t} \text{ so that}$$

$$K = \frac{P_0}{A_L \left\{E_0 + E_1 (i\omega)^{\alpha-1}\right\} H'}.$$

In conclusion, the equations for acoustic motions in rods, channels, ducts, or horns have been presented for infinitesimal longitudinal displacements in a visco-elastic. Illustration with a power law cross-sectional area, exponential, and Gaussian profiles obtains solutions for application to real materials with complex fractional calculus rheological characteristics. This offers a generalized Webster horn equation for a variety of acoustical classes of materials for experimental verification. Practical example applications include amplification, in tool design, or for absorption in vibration damping.

References

Webster, A. G. *Proceedings of the National Academy of Sciences* 5 (1919): 275-282.

Campos, L. M. B. C. *Reviews of Modern Physics* 58 (1987): 117-182.

McLachlan, N. W. *Loudspeakers.* New York: Oxford, 1934.

Graff, Karl F. *Wave Motion in Elastic Solids.* New York: Oxford, 1975.
Campos, L. M. B. C. *Physical Acoustics*. Edited by O. Leroy and M. A. Breazeale, Plenum Press, New York, 261-269, 1991.
Rouse, P. E., and K. Sittel. *Journal of Applied Physics* 24, no. 6 (1953): 690.
Bagley, R. L. *Journal of Rheology* 27, no.3 (1983): 201.
Bagley, R. L., and P. J. Torvik. *Journal of Rheology* 30, no.1,(1986): 133.
Oldham, K. B., and J. Spanier. *The Fractional Calculus*. New York: Academic Press, 2002.
McLachlan, N. W. *Bessel Functions for Engineers.* London: Oxford, 1955.

(*Exponential Horn/Displacement Amplifier/ Absorber (Visco-elastic Rheology Model) *)

```
Clear [y, a, b, c, e0, e1, areachg, f, L, frac, omega, p, rho, a0]
f = 20000
p = .528
omega = (2 Pi f )//N
cee = 1450
L = 2 cee /f+a0
frac = (I^(p))*(omega^(p))//N
e0 = 8.14* 10^5
e1 = 7.31* 10^4
rho = 1100
a0 = .003
area[x_]:= a0 Exp[2 x/ L]
areadiff[x_]:=D[area[x],x]
b[x_]:=areadiff[x]/a[x]
b = 2/L
c = omega^2/((e0/rho){1+(e1/e0) *frac })//N
aL = area[L]
p0 = 10^6
s11= Plot[Evaluate[area[x]],{x,a0, L+a0},
PlotRange->All,PlotStyle->{RGBColor[0,1,1],Thickness[0.05]},AxesL
    abel->{"Distance", "Area = a0 Exp[2 x / L]"}]
m1 = 1/L{-1+(1-c L^2)^0.5}
a1 = p0/(aL*m1*Exp[m1*L]*(e0+e1(I*omega)^(p-1)))
s2 = Plot[Evaluate[Re[a1*Exp[m1*x]],{x,a0,L+a0}],PlotRange-
    >All,PlotStyle->{RGBColor[1,0,0],Thickness[0.05]},AxesLabel-
    >{"Distance", "Re[u]"}]
s3 = Plot[Evaluate[Im[a1*Exp[m1*x]],{x,a0,L+a0}],PlotRange-
    >All,PlotStyle->{RGBColor[1,0,1],Thickness[0.05]},AxesLabel-
    >{"Distance", "Im[u]"}]
```

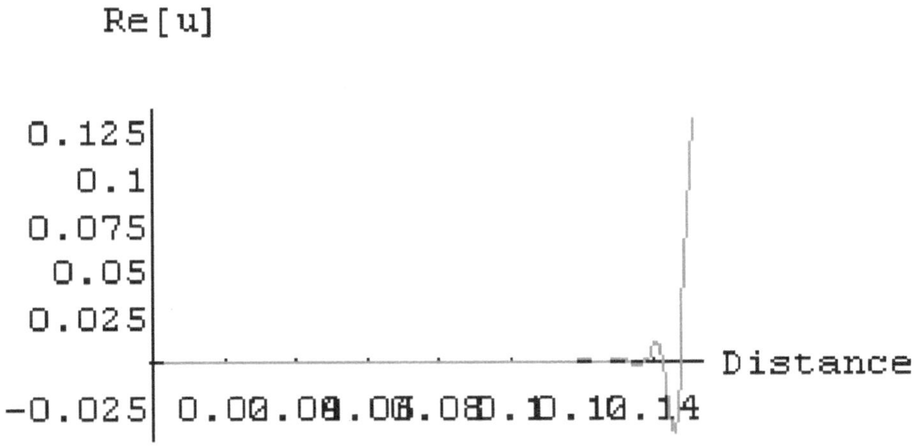

Area = a0 Exp[2 x / L]

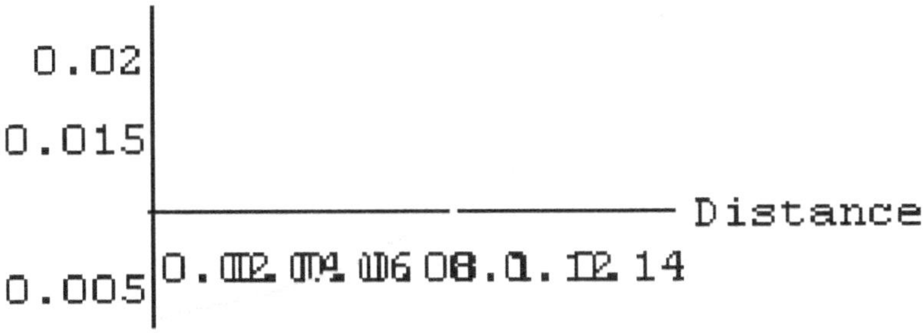

VI

Waves Through Visco-Thermal Mixtures with Multiple Chemical Reactions: Sono-Chemical Kinetics and Thermodynamic Analysis in Visco-Elastics

Abstract

Acoustic wave properties of fluids have provided fruitful qualitative and quantitative analysis of fast chemical reactions. The purpose of this paper is to present a continuum mixture acoustic theory that includes simultaneous chemical kinetic relaxation for first-order acoustic motions. The energy equations, equation of state with thermodynamic relations, are formulated to account for simultaneous chemical reactions. This sono-chemical analysis generates a general wave equation for a compressible visco-elastic with thermal transfer and for chemical progress. Longitudinal wave propagation is described by a complex biquadratic equation for the sound absorption and dispersion for arbitrary frequency but small enough amplitude for linearization yet valid continuum thermodynamic assumptions. Approximations of small dimensionless frequency numbers yield linearly additive contributions of viscous, of thermal, and of reaction effects. Here, linearly viscous or viscoelastic effects can be examined in addition to the chemical kinetic effects to the observables of sound attenuation and sound speed. Visco-elastic shear wave propagation is an outcome of this derivation. In addition, a plane wave analysis using a fractional calculus representation of the stress response versus strain and strain rate is given. Several matrix calculations are illustrated with a three-reaction example.

List of Symbols

Aj	chemical affinity of j^{-th} reaction $[ML^2T^{-2}]$
a_α	activity of α-th component [1]
B^ω_p	frequency-dependent isobaric coefficient of expansion [1]
B^ω_θ	frequency-dependent isothermal coefficient of expansion [1]
B*	complex "acoustic viscosity" $[L^2T^{-1}]$
$\overline{B}$ *	complex "acoustic viscosity" divided by density $[M^{-1}L^3T^{-1}]$
c_0, c	reference and frequency-dependent speeds of sound $[LT^{-1}]$
c_p, c_p^e	instantaneous and equilibrium heat capacities at constant pressure $[L^2T^{-2}\theta]$
c_v, c_v^e	instantaneous and equilibrium heat capacities at constant volume $[L^2T^{-2}\theta]$
C_p^ω	frequency-dependent heat capacity [1]
C_α	chemical constituent α [1]
f	frequency of wave $[T^{-1}]$
h_σ	heat of σ^{-th} reaction $[H\ mol^{-1}]$
H_σ	heat of orthonormal σ^{-th} reaction $[H\ mol^{-1}]$
k	number of constituents [1]
K_σ	equilibrium chemical rate functions σ^{-th} reaction [1]
$k^F_{\sigma,}$ k^R_σ	forward and reverse chemical rate functions σ^{-th} reaction $[M^{-1}T^{-1}]$
O( )	Order symbol
p, p^a	thermodynamic and acoustic pressures $[ML^{-1}T^{-2}]$
r	number of independent reactions r = Rank $(S_{j\alpha})$ [1]
R	universal gas constant $[ML^2T^{-2}\ \theta^{-1}mol^{-1}]$
$\underline{\mathbf{S}}$, $\mathbf{S}_{ij}$, $\underline{\mathbf{T}}$, T_{ij}	Total and extra stress tensors and components $[ML^{-1}T^{-2}]$
$S_{j\alpha}$	Signed stoichiometric coefficients
T	time variable [T]
v_σ	volume change of σ^{-th} reaction at constant temperature and pressure $[L^3\ mol^{-1}]$
V_σ	volume change of orthonormal σ^{-th} reaction $[L^3\ mol^{-1}]$
x_α mole	fraction of species α [1]
X	complex propagation function fo attenuation and dispersion $[L^{-1}]$
X	Viscosity Frequency Number [1]
Y	Thermoviscous number [1]
v, v_i	velocity vector and components i $[LT^{-1}]$
Z_σ	Extent of reaction of orthonormal σ^{-th} reaction $[mol\ M^{-1}]$
α	sound absorption $[L^{-1}]$

$\beta_{\theta,\zeta}$	instantaneous isothermal coefficient of thermal expansion $[\theta^{-1}]$
$\beta_{p,\zeta}$	instantaneous isobaric coefficient of compressibility $[T^2M^{-1}L^{-1}]$
θ, θ^a	absolute and acoustic temperatures $[\theta]$
ρ, ρ^a	Total and acoustic density $[ML^{-3}]$
μ_α	chemical potential function for $\alpha^{\text{-th}}$ component $[L^2T^{-2}mol^{-1}]$
γ	ratio of specific heats c_p / c_v [1]
γ_α	Non-ideal coefficients for $\alpha^{\text{-th}}$ species [1]
ζ_j	reaction progress variable for $j^{\text{-th}}$ reaction $[mol\ M^{-1}]$
ζ^{+e}_{j}	equilibrium reaction velocity variable for $j^{\text{-th}}$ reaction $[mol\ M^{-1}T^{1}]$
η, η_B	complex conjugate shear and bulk viscosities $[ML^{-1}T^{1}]$
λ	wavelength [L]
λ_σ	eigenvalue of $\sigma^{\text{-th}}$ reaction = 1 / τ_σ $[T^{-1}]$
v	specific volume = $1/\rho$ $[L^3M^{-3}]$
ω	angular frequency = $2\pi f$ $[T^{-1}]$
$\omega_{\alpha,}$ ω^e_α	chemical composition for $\alpha^{\text{-th}}$ component, equilibrium value $[gmol\ M^{-1}]$
τ_σ	relaxation time of orthonormal $\sigma^{\text{-th}}$ reaction [T]

The symbol [ ] means "dimensions of" {1} means "dimensionless" quantity; M = mass, L = Length, T = time, θ = temperature, H = heat (cal), mol = gmol.

$(p)' \equiv \partial p/\partial t$ = differential of p with respect to t, and
Σ represents summation; and Π denotes multiplication.

I. Introduction

The theoretical prediction of acoustic wave propagation in fluids is important for measuring physicochemical properties including rapid chemical reaction kinetics. The purpose of this paper is to present a dynamical theory applicable to a wide variety of materials, such as polymer solutions or biological systems that can exhibit non-Newtonian viscous behavior or memory effects as well as chemical reactions. The analysis utilizes field equations of continuum mechanics for homogeneous mixtures. Here the linear acoustics problem is emphasized, relying on the first-order terms. The linear equations can be solved exactly for an arbitrary frequency wave of infinitesimal amplitude with all coefficients evaluated at their uniform equilibrium reference values. A multiple timescale perturbative approach applied previously to investigate finite amplitude sound propagation in visco-elastics with second-order stress relaxation effects assuming a single chemical reaction is present. Initial acoustic studies of single-reaction liquid crystal mixtures classified as nematics have also been made.

The paper is arranged as follows: First, the chemical kinetics and thermodynamics are introduced to derive the chemical kinetic equations for forced perturbations. Then the total mass, linear momentum, and energy balance equations are used to develop a set of equations to be solved. Finally, the results are summarized in the form of a determinant equation which when expanded is a biquadratic in terms of a complex propagation variable comprised of sound absorption and sound speed information

II. Multiple Chemical Reaction Kinetics and Thermodynamics

A general reaction scheme composed of an arbitrary number of reaction steps for transient state analysis of constituents $\alpha = 1, k$ is considered. Asserting that the atomic substances are indestructible, the constituents in the mixture have been shown to follow a law of definite proportions originally experimentally based on weight analysis. The stoichiometric matrix consists of positive values for the reaction products while negative values correspond to reactants. A neutral species such as an inert solvent would be assigned a value of zero. The possible number of independent reactions r, that is, the independent rows or columns of the stoichiometric matrix equals its rank r. The concepts of linear algebra have been applied to chemical kinetics by a number of investigators. In summary, the chemical system investigated comprises,

$$\sum_{\alpha=1,k} S_{j\alpha} C_\alpha = 0 \quad \text{Independent Chemical Reactions } j = 1, r$$

$$\text{or} \sum_{\alpha=1,kR} S^-_{j\alpha} C_\alpha = \sum_{a=kR+1,k} S^+_{j\alpha} C_\alpha$$

Constituents C_α, $\alpha = 1, k$
Reactants $\alpha = 1, kR$, Reaction Products $\alpha = kR+1, k$
Signed Stoichiometric Coefficients $S_{j\alpha} = S^-_{j\alpha} - S^+_{j\alpha}$

$$\text{Sign}\left[S^{-}_{j\alpha}\right] < 0 \text{ for reactants}$$
$$\text{Sign}\left[S^{+}_{j\alpha}\right] > 0 \text{ for products}$$
$$S_{j\alpha} = 0 \text{ for an inert constituent}$$

Each part of the material body of the mixture is composed of chemical components or species α. The independent set of chemical reactions is a subset of the total number of reactions where the other dependent ones can be expressed as linear combinations of the subset chosen to describe the reaction space. The row rank r of the stoichiometric matrix, written with reactions as rows, and constituents as columns determine this independent set. In general, the column rank would differ. If the stoichiometric matrix were transposed, the column rank would have the same rank for the independent set of reactions.

The equations of mass balance for each species (i.e., sometimes called the partial mass balance equations or simply chemical kinetic equations for the mixture) may be written for the case when diffusion can be neglected as applicable to many liquid mixtures at ordinary temperatures and pressures by introducing degree of advancement or reaction progress variables, ζ_j,

$$\omega_\alpha - \omega^e_\alpha = \sum_{j=1}^{r} S_{j\alpha}\,\zeta_j$$

Chemical composition is described by [ω_α] or gmol/mass; the chemical kinetic equations set using compositions is usually larger than one using reaction progress or reaction velocity for kinetic analysis:

$$\zeta'_j = \zeta^+_j$$

Assuming a polynomial mass-action form for the kinetics' constitutive relation or production term ζ^+_j on the right-hand side of the equation above,

$$\zeta'_j = k^F_j(\theta, p)\prod_{\alpha\ \mathbf{reac\tan ts}} a_\alpha^{S^-_{j\alpha}} - k^R_j(\theta, p)\prod_{\alpha\ \mathbf{products}} a_\alpha^{S^+_{j\alpha}}$$

All composition changes may be expressed in terms of these degree of advancement variables using the chosen r independent reactions. The large font pi symbols denote multiplication, and the superscript accent on the reaction variable denotes a time differentiation. The concentrations are mathematical bases raised to their appropriate stoichiometric coefficient for that reaction. The functions k^F_j and k^R_j are the forward and reverse reaction rate coefficients dependent on temperature and on pressure that are assumed independent of the activities.

At chemical equilibrium with no external sources, the chemical production vanishes $\zeta^+_j = 0$, yielding the equilibrium functions for the chemical system understudy. The expression $k^F_j / k^R_j = K_j(\theta, p) = \Pi_{\alpha=1,k} a_\alpha{}^{Sj\alpha} = \exp(A_j^\varnothing / R\theta)$ defines the equilibrium functions of both temperature and of pressure, where $A_j^\varnothing$ is the standard affinity of the jth reaction. The chemical affinity introduced by de Donder is conveniently defined in terms of the chemical potentials $\mu_{\alpha,}$

$$A_j(\theta, p, \zeta) = -\Sigma_{\alpha=1,k} S_{j\alpha}\, \mu_\alpha \text{ with } \mu_\alpha = \mu^\varnothing_\alpha(\theta, p) + R\,\theta\, ln\, a_\alpha$$

and activities are written $a_\alpha = x_\alpha \gamma_\alpha$. γ_α are the non-ideality coefficients for the mole fraction of component α denoted by x_α. Perturbations of chemical affinities $A_j(\theta, p, \zeta_j)$ for the j independent reactions as a function of temperature θ and pressure p derive a set of linear chemical kinetic equations written as,

$$\zeta^{a\prime}_j = \zeta^{+e}_j \{ \{ \partial [A_j / R\theta] / \partial\theta \} \theta^a + \{ \partial [A_j / R\theta] / \partial p \} p^a + \Sigma_{s=1,r} \{ \partial [A_j / R\theta] / \partial \zeta_s \} \zeta_s{}^a \}$$

The following thermodynamic definitions are incorporated in the analysis:

$$\partial [A_j / (R\theta)] / \partial\theta = h_j / (R\theta^2),\ \partial [A_j / (R\theta)] / \partial p = -v_j / (R\theta).$$

For $\theta^a = p^a = 0$, the chemical kinetic analysis obtains a linear system of coupled ordinary differential equations of first order in ζ^a_j. It is convenient to define a new set of base vectors whose elements are formed from the using the determinant equation

$$\det \left| \underline{M}^{-1}\, \underline{J}\, \underline{M} - \lambda\, \underline{I} \right| = 0$$

Here $\partial [A_j / [R\theta]] / \partial\zeta_s \equiv a_{js}$, $J_{js} = \Sigma_{j's=1,r} \{ \zeta^{+e}_j \}^{0.5} a_{js} \{ \zeta^{+e}_j \}^{0.5} = J_{sj}$. Let $\underline{M}$ be a matrix constructed from the r eigenvectors corresponding to the eigenvalues of the matrix $\underline{J}$; a_{js} and J_{js} are symmetric matrices, which yield real eigenvalues λ. The matrix $\underline{I}$ is the identity matrix. Now the chemical kinetic equations may be cast into independent disjoint equations of normal form.

The reciprocal chemical relaxation time for reaction j, $1 / \tau_j$, equals the eigenvalue λ_j. Eigenvalue determinations by direct or approximate methods for illustrative chemical systems were previously discussed. Refer to appendix B.

Transforming to orthonormal reaction velocity or extent variables

$$\zeta^a_j = \{ \zeta^{+e}_j \}^{0.5} \Sigma_{s=1,r} M_{js} Z^a_s \text{ where}$$
$$\underline{M}^{-1}\underline{J}\,\underline{M} = \underline{\Lambda} \text{ (diagonal matrix of } \lambda_j \text{) or } \Sigma_{s,\tau=1,r} M^{-1}_{sj} J_{j\tau} M_{\tau v} = \lambda_{jv}\, \delta_{jv}$$
$$\text{Then } \underline{M}^T \underline{M} = \underline{I} = \underline{M}\,\underline{M}^T$$

Acoustic quantities representing perturbations about equilibrium: $p^a = p - p^e$, $\theta^a = \theta - \theta^e$, and $\zeta^a = \zeta - \zeta^e$ are used to express the chemical kinetic rate equations or balances below,

$$Z^{a\prime}_{\underline{j}} = \{ H_{\underline{j}} / (R\, \theta^2) \}\, \theta^a - \{ V_{\underline{j}} / (R\, \theta) \}\, p^a - \{ 1 / \tau_{\underline{j}} \}\, Z^a_{\underline{j}}$$

(The underlined subscripts denote that no summation is to be taken).

The transformed thermodynamic parameters become,

$$H_j = \Sigma_{s=1,r} M_{js} \{ \zeta^{+e}_s \}^{0.5} h_s \, , \; V_j = \Sigma_{s=1,r} M_{js} \{ \zeta^{+e}_s \}^{0.5} v_s .$$

The analysis pertains to forced harmonic waves which motivate introducing,

$$Z^a_{\underline{j}} = Z^0_{\underline{j}} \exp [i\, \omega\, t] \text{ so that } Z^{a\prime}_{\underline{j}} = i\, \omega\, Z^a_{\underline{j}} \text{ and}$$

$$Z^a_{\underline{j}} = \{ H_{\underline{j}}\, \tau_{\underline{j}} / [R\theta^2 (1 + i\, \omega\, \tau_{\underline{j}})] \}\, \theta^a - \{ V_{\underline{j}}\, \tau_{\underline{j}} / [R\, \theta\, (1 + i\, \omega\, \tau_{\underline{j}})] \}\, p^a \quad (1)$$

Additionally, the same harmonic variation is made for temperature and for pressure and will be implemented with the other linear momentum and energy equations.

III. Balances of Total Mass, Linear Momentum, and Energy

A. Total mixture mass

The balance of total mass yields for no external sources:

$$\partial\rho/\partial t + \nabla \bullet (\rho\, \mathbf{v}) = 0$$

Acoustic perturbations may also be written as,

$$\rho = \rho_0 + \rho^a,\ \mathbf{v} = \mathbf{v}_0 + \mathbf{v}^a, \text{ with } \mathbf{v}_0 = 0 \text{ for a fluid at rest,}$$
$$O(1):\quad \partial\rho^a/\partial t + \rho_0 \nabla \bullet (\mathbf{v}^a) = 0$$

The density function $\rho(\theta, p, \underline{Z})$ is expanded by a Taylor series about an equilibrium position (where strong chemical equilibrium is assumed):

$$\rho^a = (\rho,_\theta)\,\theta^a + (\rho,_p)\,p^a + \Sigma_{j,s=1,r} \{\partial\rho / \partial\zeta_j\} \{\partial\zeta_j / \partial Z_s\} Z^a_s.$$

Further thermostatic definitions are shown as follows,

$$\partial\rho / \partial\theta \equiv \rho,_\theta = -\beta_{\theta,\zeta}/\upsilon_0\, \rho,_p = \beta_{p,\zeta}/\upsilon_0$$

$$\{\partial\rho / \partial\zeta_j\} \{\partial\zeta_j / \partial Z_s\} = -\rho_0\, \Sigma_{j=1,r} \{\zeta^{+e}_j\}^{0.5} \nu_j M_{js}$$

Substituting the density function approximation into the balance of total mass yields,

$$-\{\beta_{\theta,\zeta}/\upsilon_0\}\,\partial\theta^a/\partial t + \{\beta_{p,\zeta}/\upsilon_0\}\,\partial p^a/\partial t$$
$$-\Sigma_{j,s=1,r}\, \rho_0 \{\zeta^{+e}_j\}^{0.5} \nu_j M_{js} \{Z_s^a/\partial t\} + \rho_0 \nabla \bullet (\mathbf{v}^a) = 0$$

(2)

Combining Eq. (1) and Eq. (2) obtains,

$$-\beta_{\theta,\zeta} B^\omega_\theta \{\partial\theta^a/\partial t\} + \beta_{p,\zeta} B^\omega_p \{\partial p^a/\partial t\} + \nabla \bullet (\mathbf{v}^a) = 0 \qquad (3)$$

Also, it is noted,

$$B^{\omega}_{\theta} = 1 + \{ (H_j V_j \tau_j) / [\upsilon_0 \beta_\theta R \theta_0{}^2 (1 + i \omega \tau_j)]\}$$
$$= 1 + \Sigma_j \{ \Delta B_{\theta p} / (1 + i \omega \tau_j)\}$$
$$B^{\omega}_{p} = 1 + \{ V_j^2 \tau_j / [\upsilon_0 \beta_p R \theta_0 (1 + i \omega \tau_j)]\}$$
$$= 1 + \Sigma_j \{ \Delta B_{pp} / (1 + i \omega \tau_j)\}$$

B. Linear momentum

The balance of total linear momentum is shown as,

$$\rho (\partial \mathbf{v} / \partial t + \mathbf{v} \bullet \nabla \mathbf{v}) = \nabla \bullet \underline{S} + \rho \mathbf{F}_b$$

with total stress tensor $\underline{S} = - p \underline{I} + \underline{T}$ and extra stress $\underline{T}$

$$\underline{T} = \int_{-\infty} \{(K(s) - 2/3 G(s)) \operatorname{tr} \underline{D}(\tau) \} \underline{I} d\tau + 2 \int_{-\infty} G(s) \underline{D}(\tau) d\tau$$

$\underline{D}$ represents the symmetric part of the velocity gradient or strain rate tensor and with a body force vector $\mathbf{F}_b$ which is presently neglected.

The upper integration limit is the present time t; past time $s = t - \tau$.

The stress tensor represents a relatively large class of nonlinear fluids with memory to first-order. A fractional calculus rheology model for polymers is shown in the appendix A.

Using a vector identity $\nabla\times (\nabla\times \mathbf{v}) = \nabla(\nabla \bullet \mathbf{v}) - \nabla^2 \mathbf{v}$ obtains

$$\rho (\partial \mathbf{v} / \partial t + \mathbf{v} \bullet \nabla \mathbf{v}) = - \nabla p + \int_{-\infty}^{t} \{ (K(s) + 4/3 G(s)) \nabla (\nabla \bullet \mathbf{v})\} d\tau +$$
$$- \int_{-\infty}^{t} G(s) \nabla\times (\nabla\times \mathbf{v}) d\tau]$$

O(1): ____________________

$$\rho_0\, \partial \mathbf{v}^a / \partial t. = - \nabla p^a + \int_{-\infty}^{t} \{K(s) + 4/3\, G(s)\}\, \nabla (\nabla \bullet \mathbf{v}^a)\, d\tau$$

$$- \int_{-\infty}^{t} G(s)\, \nabla\times (\nabla\times \mathbf{v}^a)\, d\tau \qquad (4)$$

Differentiating Eq. (4) with respect to time (using superscript accent notation) and subtracting $c_0^{\,2} / \gamma \nabla (\nabla \bullet \mathbf{v}^a)$ yields a wave equation formulation sought

---

$$\partial^2 \mathbf{v}^a / \partial t^2 - c_0^{\,2} / \gamma \nabla (\nabla \bullet \mathbf{v}^a)$$

$$= - \nabla p^{a\prime} / \rho_0 + \int_{-\infty}^{t} [\{K(s) + 4/3\, G(s)\} / \rho_0]\, \nabla (\nabla \bullet \mathbf{v}^{a\prime})\, d\tau$$

$$- \int_{-\infty}^{t} G(s)\, \nabla\times (\nabla\times \mathbf{v}^{a\prime})\, d\tau - c_0^{\,2} / \gamma \nabla (\nabla \bullet \mathbf{v}^a) \qquad (5)$$

C. Energy

The equation for the internal energy, which represents the difference of the total and kinetic energies is written assuming Fourier heat conduction (i.e., heat flux is proportional to temperature gradient) and strong chemical equilibrium (vanishing of chemical affinities). A derivation follows for this reacting fluid system to first-order O(1)

$$\rho_0 c_P \frac{\partial \theta^a}{\partial t} - \theta_0 \beta_\theta \frac{\partial p^a}{\partial t} + \rho_0 h_{\underline{\sigma}} \left(\varsigma_{\underline{\sigma}}^{+e}\right)^{0.5} M_{\sigma s} \frac{\partial Z_s^a}{\partial t} - k_\theta \nabla^2 \theta^a = 0$$

Rewriting the third term as $\rho_0 \sum_j H_j Z_j^{a'}$ obtains

$$\rho_0 c_P \frac{\partial \theta^a}{\partial t} - \theta_0 \beta_\theta \frac{\partial p^a}{\partial t} + \rho_0 \sum_j H_j \frac{\partial Z_s^a}{\partial t} - k_\theta \nabla^2 \theta^a = 0 \ .$$

Next, substitute the previously derived chemical rate expression that has been linearized for harmonic variation to yield

$$\rho_0 c_P \frac{\partial \theta^a}{\partial t} - \theta_0 \beta_\theta \frac{\partial p^a}{\partial t} + \rho_0 \sum_j H_j \left(\frac{H_j \tau_j}{R\theta_0^2 (1 + i\omega\tau_j)} \frac{\partial \theta^a}{\partial t} - \frac{V_j \tau_j}{R\theta_0^2 (1 + i\omega\tau_j)} \frac{\partial p^a}{\partial t} \right) - k_\theta \nabla^2 \theta^a = 0$$

Collecting terms,

$$\rho_0 (c_P + \sum_j \frac{H_j^2 \tau_j}{R\theta_0^2 (1 + i\omega\tau_j)}) \frac{\partial \theta^a}{\partial t} - \theta_0 \beta_\theta \left(1 + \sum_j \frac{H_j V_j \tau_j}{R\theta_0^2 (1 + i\omega\tau_j)} \right) \frac{\partial p^a}{\partial t} - k_\theta \nabla^2 \theta^a$$

Therefore to O(1) the energy equation incorporating reaction with accompanying density (or volume) change simplifies into a compact form:

$$\rho_0 (c_P C_P^{\omega}) \frac{\partial \theta^a}{\partial t} - \theta_0 \beta_\theta B_\theta^{\omega} \frac{\partial p^a}{\partial t} - k_\theta \nabla^2 \theta^a = 0 \quad \ldots \quad (6)$$

The linear acoustic equations, admitting shear wave motions, proceeds as follows. Introduce scalar and vector potentials,

$\mathbf{v}^a = - \nabla\varphi + \nabla\times \mathbf{A}$, with sinusoidal periodic motions,
$\mathbf{v}^a = \mathbf{v}^0 \exp[i\,\omega\, t]$, $p^a = p^0 \exp[i\,\omega\, t]$,
$\nabla \bullet \mathbf{A} = 0$, writes Eq. (5) as,

$$\omega^2 \nabla \phi + i\,\omega \nabla p^a / \rho_0 + i\,\omega\, \overline{B}* \nabla^2 \varphi = - \nabla \times \{ \omega^2 \mathbf{A} + G^*/ \rho_0 \nabla^2 \mathbf{A} \} \quad (7)$$

Let the acoustic viscosity/density term be defined as, $\overline{B}* \equiv \{K^* + 4/3\, G^*\} / \rho_0$. This linear combination of viscosities (divided by the density) is sought by acoustic measurement. $K^* = i\,\omega\,\eta_B$, $G^* = i\,\omega\,\eta$, (G^* and K^* are written in complex conjugate form as previously derived). G^* and K^* capture elastic or viscous relaxation but account for the temperature in a classical way in that the stress tensor depends on the present value of temperature only. This assumption also pertains to the internal energy in the energy equation. Special cases of elastic Newtonian fluid and a wide variety of non-Newtonian materials are represented by these functions.

Shear wave motions are described by,

$$\nabla^2 \mathbf{A} + \kappa^2 \mathbf{A} = 0 \quad \kappa^2 = \rho_0 \, \omega^2 / G^* \tag{8}$$

D. Biquadratic characteristic equation

The O(1) motions form a set of algebraic equations when linear harmonic motions are assumed, which are relevant for many experiments and commercial applications. This requires the following determinant equation to be solved for the complex propagation coefficient X comprising the attenuation and sound velocity observables. The determinant is written as follows for the independent variables θ^a, p^a, and ϕ^a after multiplying the first and third rows by (-1) and the second row by 1/X.

$$\det\begin{pmatrix} i\omega\beta_{\theta,\varsigma}B_\theta^\omega & -i\omega\beta_{p,\varsigma}B_p^\omega & X^2 \\ 0 & \dfrac{i\omega}{\rho_0} & \omega^2 + i\omega\overline{B}^* X^2 \\ -i\omega\rho_0 c_p C_p^\omega + k_\theta X^2 & i\omega\theta_0\beta_{\theta,\varsigma}B_\theta^\omega & 0 \end{pmatrix} = 0$$

Expanding yields a biquadratic, a $W^4 + b\,W^2 + c = 0$, where $W \equiv k_0 / X$ and $X = -(\alpha + i\,\omega / c)$, the complex acoustic propagation coefficient whose real part contains the attenuation information and the imaginary part contains the dispersion.

$$a = B_p^\omega C_p^\omega + \frac{\gamma(\gamma-1)B_p^{\omega 2}B_\theta^{\omega 2}\omega^2}{\rho_0^2\beta_\theta^2 c_0^4},$$

$$b = \frac{C_p^\omega}{\gamma} + iB_p^\omega XY + iB_p^\omega C_p^\omega X + \frac{i\gamma(\gamma-1)B_p^{\omega 2}B_\theta^{\omega 2}X\omega^2}{\rho_0^2\beta_\theta^2 c_0^4}$$

$$c = \frac{iXY}{\gamma} - B_p^\omega X^2 Y \quad ; \quad X \equiv \frac{\overline{B}^*\omega}{c_0^2} \quad Y \equiv \frac{k_t}{\overline{B}^* c_0^2}$$

$$k_t = \frac{k_\theta}{\rho_0 c_p}$$

Dimensionless quantities are referred to as the frequency number X and the thermoviscous number Y.

The classical Kirchhoff-Langevin biquadratic solution for nonreacting cases has been modified to include multiple independent orthonormal chemical reactions. An earlier partitioned determinant equation approach yields the same results with the restriction to plane waves without a shear wave contribution. A fractional calculus rheology model is addressed in the appendix I. The reactive single reaction case for a Newtonian fluid and non-Newtonian materials has been extended to include multiple chemical reactions that also admits shear waves.

E. Expansion in low frequency viscosity number

Writing part of a Taylor series expansion in viscosity frequency number X and thermal conduction parameter Y gives formulas for practical application:

$$W^2 \equiv P; \text{ therefore } a\,P^2 + b\,P + c = 0;$$

$$P = P_0 + (\partial P / \partial X)\big|_{X=0} X + (\partial P / \partial Y)\big|_{Y=0} Y$$

$$\text{for } P_0 = -\,c / b\,\big|_{X=0} \text{ and}$$

$$(\partial P / \partial X)\big|_{X=0} = [\,-b_X P_0 - a_X P_0^{\,2}\,] / b\,\big|_{X=0}$$

V. Conclusion

The paper achieved its purpose of deriving a mathematical continuum acoustic theory for the problems of acoustic wave propagation in a wide variety of viscous materials with simultaneous chemical reactions such as polymers that exhibit memory effects. The sono-chemical kinetic analysis is important for studying multiple-step, fast chemical reactions (stoichiometry, heats, and volume changes) as a kind of acoustic spectrometry for measuring viscosity (including bulk or compressible viscosity), for examining thermodynamic equations of state and solvation, for developing nanotechnology applications, for assessing environmental sound transmission such as through seawater, as well as for biochemical enzyme, and for chemical technology monitoring and analysis. Further basic research is recommended to explore and understand alternative non-mass-action chemical kinetic rate equation constitutive models with fractal chemical diffusive rates, possible role of a shear-dependent reaction rate term, as well as thermal transport memory effects.

Acknowledgment

The author wishes to thank Prof. William H. Schwarz for suggesting the research and Prof. Robert Green at Johns Hopkins University for his advice on the subject of this paper.

Appendix A: Fractional Calculus Rheology Model

Converting acoustic velocities to time differentials of u^a, denoting displacement and perturbing harmonically for the extra stress given by

$$T = E_0 \frac{\partial u^a}{\partial x} + E_1 \frac{\partial}{\partial x} \frac{\partial^\alpha}{\partial t^\alpha} \frac{\partial u^a}{\partial t}$$

helps formulate a variation of the problem given in the body of the paper.

This fractional calculus stress for elastico-viscous response applies to many polymer solutions with $\alpha = 0.5$ (i.e., sometimes referred to as semi-differentials) thereby simplifying the previous expressions with a larger number of terms for material stress response representation in several cases examined. Expanding the new determinant yields solutions for this fractional calculus material representation. The fractional derivative α has been found useful in many applications:

$$\frac{\partial^\alpha f}{\partial t^\alpha} \equiv \frac{1}{\Gamma(1-\alpha)} \frac{d}{dt} \int_0^t \frac{f(\tau)d\tau}{(t-\tau)^\alpha}, \; 0 < \alpha < 1.$$

From a Fourier transform approach using an integration of parts with vanishing of the function at $\pm\infty$ has given $\frac{\partial^\alpha}{\partial t^\alpha}\left(e^{i\omega t}\right) = (i\omega)^\alpha \left(e^{i\omega t}\right)$. More generally,

$$\frac{\partial^\alpha}{\partial t^\alpha}\left(e^{i\omega t}\right) = (\omega)^\alpha \left(e^{i(\omega t + \frac{\pi\alpha}{2})}\right) \quad \text{for } \omega \geq 0.$$

The previous analysis in the text for a Fourier thermal conductor can be modified for a fractional calculus rheology model, rewriting the equations for the independent variables θ^a, p^a, and u^a with the abovementioned fractional differentiation of the harmonic variations assumed.

Appendix B

Consider a set of three multiple-step reactions in the forward and reverse directions. Here below is a summary of the stoichiometry matrix and the equilibrium reaction progress variables for an ideal system. In addition, the characteristic equation is provided to calculate the eigenvalues or reciprocal chemical relaxation times of the system. The corresponding eigenvectors to help form an orthonormal set of reactions are also given to help illustrate the methods and mathematics in the paper.

$$C_1 + C_2 \Leftrightarrow C_3 + C_4$$

$$C_3 + C_4 \Leftrightarrow C_5 + C_6$$

$$C_5 + C_6 \Leftrightarrow C_7 + C_8$$

$$S_{j\alpha} = \begin{pmatrix} -1 & -1 & 1 & 1 & 0 & 0 & 0 & 0 \\ 0 & 0 & -1 & -1 & 1 & 1 & 0 & 0 \\ 0 & 0 & 0 & 0 & -1 & -1 & 1 & 1 \end{pmatrix}$$

$$S_1 = S_2 = S_3 = 0$$

$$\varsigma_1^{+F} = k_1^F x_1 x_2 \quad \varsigma_1^{+R} = k_1^R x_3 x_4 \quad K_1 = \frac{x_3 x_4}{x_1 x_2}$$

$$\varsigma_2^{+F} = k_2^F x_3 x_4 \quad \varsigma_2^{+R} = k_2^R x_5 x_6 \quad K_2 = \frac{x_5 x_6}{x_3 x_4}$$

$$\varsigma_3^{+F} = k_3^F x_5 x_6 \quad \varsigma_3^{+R} = k_3^R x_7 x_8 \quad K_3 = \frac{x_7 x_8}{x_5 x_6}$$

A cubic polynomial formed from the determinant equation obtains the eigenvalues $\lambda_{1,}$ $\lambda_{2,}$ and $\lambda_{3.}$ These have also been used to define scalar invariants I.

$$\left|J_{ij} - \lambda\delta_{ij}\right| = 0 \quad \Rightarrow \quad \lambda^3 - I_1\lambda^2 + I_2\lambda - I_3 = 0$$

$$I_1 = J_{ii} = J_{11} + J_{22} + J_{33}$$

$$I_2 = \begin{vmatrix} J_{11} & J_{21} \\ J_{12} & J_{22} \end{vmatrix} + \begin{vmatrix} J_{22} & J_{32} \\ J_{23} & J_{33} \end{vmatrix} + \begin{vmatrix} J_{11} & J_{31} \\ J_{13} & J_{33} \end{vmatrix} = \frac{1}{2}\left(J_{ii}J_{jj} - J_{ij}J_{ji}\right)$$

$$I_3 = \begin{vmatrix} J_{11} & J_{12} & J_{13} \\ J_{21} & J_{22} & J_{23} \\ J_{31} & J_{32} & J_{33} \end{vmatrix}$$

$$I_1 = \lambda_1 + \lambda_2 + \lambda_3$$

$$I_2 = \lambda_1\lambda_2 + \lambda_2\lambda_3 + \lambda_3\lambda_1$$

$$I_3 = \lambda_1\lambda_2\lambda_3$$

Transforming variables enables the solution of the cubic as a quadratic; however, the sign, information, and cube roots of negative one are used to select the roots.

$$\lambda^3 - I_1\lambda^2 + I_2\lambda = 0 \qquad \text{Eq(B1)}$$

$$\text{Let } \lambda = y + \frac{I_1}{3} \qquad \text{Eq(1)} \Rightarrow y^3 + py + q = 0 \qquad \text{Eq(B2)}$$

$$p \equiv I_2 - \frac{I_1^2}{3} \qquad q \equiv \frac{I_1 I_2}{3} - \frac{2}{27} I_1^3$$

$$\text{Let } y = z - \frac{p}{3z} \qquad \text{Eq(B2)} \Rightarrow z^3 - \frac{p^3}{27z^3} + q = 0 \qquad \text{or} \qquad z^6 + qz^3 - \frac{p^3}{27} = 0$$

Solving as a quadratic equation in z^3 obtains,

$$z^3 = -\frac{q}{2} + \sqrt{R} \quad \text{and} \quad z^3 = -\frac{q}{2} - \sqrt{R} \qquad R \equiv \left(\frac{p}{3}\right)^3 + \left(\frac{q}{2}\right)^2 > 0$$

$$\text{For} \quad z_1 = \sqrt[3]{-\frac{q}{2} + \sqrt{R}}, \qquad z_2 = \sqrt[3]{-\frac{q}{2} - \sqrt{R}}, \qquad \varpi = -.5 + .5\sqrt{3}i, \qquad \varpi^2 = -0.5 - .5\sqrt{3}i$$

$$\text{since} \quad z_1 z_2 = -\frac{p}{3}, \qquad \varpi z_1 \varpi^2 z_2 = -\frac{p}{3}, \qquad \varpi^2 z_1 \varpi z_2 = -\frac{p}{3}$$

$$y_1 = z_1 + z_2 \qquad y_2 = \varpi z_1 + \varpi^2 z_2 \qquad y_3 = \varpi^2 z_1 + \varpi z_2$$

$$\text{For} \quad R < 0 \quad \text{let} \quad y = nz \quad \text{to solve} \quad y^3 + py + q = 0 \quad \Rightarrow z^3 + \frac{p}{n^2} z + \frac{q}{n^3} = 0$$

$$z = \cos A \qquad n = \sqrt{\frac{-4}{3} p} \qquad \cos(3A) = \frac{-.5q}{\sqrt{\frac{-p^3}{27}}}$$

$$y_1 = nz_1 \qquad y_2 = nz_2 \qquad y_3 = nz_3$$

$$z_1 = \cos A \qquad z_2 = \cos\left(A + 120^\circ\right) \qquad z_3 = \cos\left(A + 240^\circ\right)$$

For each eigenvalue, an eigenvector is constructed to within a constant factor by the system of equations.

$$\left(J_{ik} - \lambda_r \delta_{ik}\right) M_k^r = 0$$

As an example, consider $\lambda = \lambda_1$ and solve the system of equations to obtain relative eigenvector information, which is then used with a normalization condition to construct orthonormal vector elements.

$$\left(J_{11} - \lambda_1\right) M_1^1 + J_{12} M_2^1 + J_{13} M_3^1 = 0$$
$$J_{21} M_1^1 + \left(J_{22} - \lambda_1\right) M_2^1 + J_{23} M_3^1 = 0$$
$$J_{31} M_1^1 + J_{32} M_2^1 + \left(J_{33} - \lambda_1\right) M_3^1 = 0$$

The directions are uniquely determined for the eigenvector M^1, and the other principal directions are derived in a similar manner for M^2 and M^3.

$$\frac{M_1^1}{M_2^1} = \frac{\begin{vmatrix} J_{22} - \lambda_1 & J_{23} \\ J_{32} & J_{33} - \lambda_1 \end{vmatrix}}{\begin{vmatrix} J_{23} & J_{21} \\ J_{33} - \lambda_1 & J_{31} \end{vmatrix}} \qquad \frac{M_2^1}{M_3^1} = \frac{\begin{vmatrix} J_{23} & J_{21} \\ J_{33} - \lambda_1 & J_{31} \end{vmatrix}}{\begin{vmatrix} J_{21} & J_{22} - \lambda_1 \\ J_{31} & J_{32} \end{vmatrix}}$$

Next, apply a unit magnitude condition,

$$\left(M_1^1\right)^2+\left(M_2^1\right)^2+\left(M_3^1\right)^2=1$$

Single Reaction Example:

$\omega_\alpha - \omega_\alpha^e \equiv \sum_{j=1}^{r} S_{j\alpha}\zeta_j = S_{1\alpha}\zeta_1$ The kinetic's equation states

$\dot{\zeta}_1 = \overset{+}{\zeta}_1 = \overset{+F}{\zeta}_1 - \overset{+R}{\zeta}_1$ where

$$\omega_T = \sum_{\alpha=1}^{k}\omega_\alpha \qquad x_\alpha \equiv \frac{\omega_\alpha}{\omega_T} \qquad M_w^e = \sum_{\alpha=1}^{k} x_\alpha M_{w\alpha}$$

and $\omega_\alpha \equiv \dfrac{n_\alpha}{m_\alpha} =_\alpha \dfrac{\dfrac{m_\alpha}{M_{w\alpha}}}{m_\alpha} = \dfrac{1}{M_{w\alpha}}$

The kinetics written in terms of chemical affinities and potentials obtains,

$$\dot{\zeta}_1 = \overset{+e}{\zeta}_1\left(\frac{\partial\left(\frac{A_1}{R\theta}\right)}{\partial\zeta_1}\right)\zeta_1^a \qquad A_1 = -\sum_{\alpha=1}^{k} S_{1\alpha}\mu_\alpha \qquad \mu_\alpha = \mu_\alpha^\ominus + R\theta \ln a_\alpha \ .$$

The activities $a_\alpha = \gamma_\alpha x_\alpha$; $\gamma_\alpha = 1$ for an ideal system. Using mass-action kinetics,

$$\overset{+e}{\zeta}_1 = k_1^F(\theta, p)\prod_{\text{reactants}} a_\alpha^e = k_1^R(\theta, p)\prod_{\text{rproducts}} a_\alpha^e$$

and $\dfrac{1}{\tau_1} = \overset{+e}{\zeta}_1 \dfrac{\partial}{\partial\zeta_1}\left({}^{A_1}\!/_{R\theta}\right) = \overset{+e}{\zeta}_1\left\{\sum_{\alpha=1}^{k}\dfrac{S_\alpha^2}{\omega_\alpha} - \dfrac{S_1^2}{\omega_T^e} + \sum_{\alpha=1}^{k}\dfrac{\partial}{\partial\zeta_1}\ln\gamma_\alpha\right\}$

Reaction	ζ_1^F	ζ_1^R	$K_1(\theta, p)$	S_{11}^-	S_{21}^-	S_{31}^+	S_{41}^+	S_1
$C_1 + C_2 \leftrightarrow C_3$	$k_1^F x_1 x_2$	$k_1^R x_3$	$x_3 / x_1 x_2$	-1	-1	1	0	-1
$C_1 \leftrightarrow C_3 + C_4$	$k_1^F x_1$	$k_1^R x_3$	$x_3 x_4 / x_1$	-1	0	1	1	1
$C_1 + 2\,C_2 \leftrightarrow C_3$	k_1^F	$k_1^R x_3$	$x_3 / x_1 x_2^2$	-1	-2	1	0	-2
$C_1 + C_2 \leftrightarrow C_3 + C_4$	$k_1^F x_1 x_2$	$k_1^R x_3 x_4$	$x_3 x_4 / x_1 x_2$	-1	-1	1	1	0
$2\,C_1 \leftrightarrow C_3$	$k_1^F x_1^2$	$k_1^R x_3$	x_3 / x_1^2	-2	0	1	0	-1
$n\,C_1 \leftrightarrow C_3$	$k_1^F x_1^n$	$k_1^R x_3$	x_3 / x_1^n	-n	0	1	0	1-

Reaction	Chemical Afinity A_1	$\frac{1}{\overset{+e}{\zeta_1}\tau_1} = \left\{\sum_{\alpha=1}^{k} \frac{S_\alpha^2}{\omega_\alpha}\right\} - \frac{S_1^2}{\omega_T^e}$
$C_1 + C_2 \leftrightarrow C_3$	$-\mu_1 - \mu_2 + \mu_3$	$\frac{(-1)^2}{\omega_1^e} + \frac{(-1)^2}{\omega_2^e} + \frac{1^2}{\omega_3^e} - \frac{(S_1)^2}{\omega_T^e}$
$C_1 \leftrightarrow C_3 + C_4$	$-\mu_1 + \mu_3 + \mu_4$	$\frac{(-1)^2}{\omega_1^e} + \frac{1^2}{\omega_2^e} + \frac{1^2}{\omega_3^e} - \frac{(S_1)^2}{\omega_T^e}$
$C_1 + 2\,C_2 \leftrightarrow C_3$	$-\mu_1 - 2\mu_2 + \mu_3$	$\frac{(-1)^2}{\omega_1^e} + \frac{(-2)^2}{\omega_2^e} + \frac{1^2}{\omega_3^e} - \frac{(S_1)^2}{\omega_T^e}$
$C_1 + C_2 \leftrightarrow C_3 + C_4$	$-\mu_1 - \mu_2 + \mu_3 + \mu_4$	$\frac{(-1)^2}{\omega_1^e} + \frac{(-1)^2}{\omega_2^e} + \frac{1^2}{\omega_3^e} + \frac{1^2}{\omega_4^e} - \frac{(S_1)^2}{\omega_T^e}$
$2\,C_1 \leftrightarrow C_3$	$-2\mu_1 + \mu_3$	$\frac{(-2)^2}{\omega_1^e} + \frac{1^2}{\omega_3^e} - \frac{(S_1)^2}{\omega_T^e}$
$n\,C_1 \leftrightarrow C_3$	$-n\mu_1 + \mu_3$	$\frac{(-n)^2}{\omega_1^e} + \frac{1^2}{\omega_3^e} - \frac{(S_1)^2}{\omega_T^e}$

Two Simultaneous Independent Reactions:

$$S_{11}^{-}C_1 + S_{12}^{-}C_2 \Leftrightarrow S_{13}^{+}C_3$$
$$S_{21}^{-}C_1 + S_{22}^{-}C_2 + S_{23}^{-}C_3 \Leftrightarrow S_{24}^{+}C_4 + S_{25}^{+}C_5$$

Stoichiometric Matrix: $S_{j\alpha} = \begin{pmatrix} -S_{11}^{-} & -S_{12}^{-} & -S_{13}^{+} & 0 & 0 \\ -S_{21}^{-} & -S_{22}^{-} & -S_{23}^{-} & S_{24}^{+} & S_{25}^{+} \end{pmatrix}$

$$S_{j=1,2} = \sum_{\alpha=1}^{k} S_{j\alpha} \qquad S_1 = \left(-S_{11}^{-} - S_{12}^{-} + S_{13}^{+}\right) \qquad S_2 = \left(-S_{21}^{-} - S_{22}^{-} + S_{24}^{+}\right) + S_{25}^{+}$$

$$\begin{pmatrix} \dot{\zeta}_1^{a} \\ \dot{\zeta}_2^{a} \end{pmatrix}^{\bullet} = \begin{pmatrix} \overset{+}{\zeta}_1^{e} A_{11} & \overset{+}{\zeta}_2^{e} A_{12} \\ \overset{+}{\zeta}_2^{e} A_{21} & \overset{+}{\zeta}_2^{e} A_{22} \end{pmatrix} \begin{pmatrix} \dot{\zeta}_1^{a} \\ \dot{\zeta}_2^{a} \end{pmatrix} \text{ where } \overset{+}{\zeta}_1^{e} = k_1^{F}(\theta, p)\, x_1^{S_{11}^{-}} x_2^{S_{12}^{-}} = k_1^{R}(\theta, p)\, x_3^{S_{13}^{+}}$$

$$\overset{+}{\zeta}_2^{e} = k_2^{F}(\theta, p)\, x_1^{S_{21}^{-}} x_2^{S_{22}^{-}} x_3^{S_{23}^{-}} \quad = \quad k_2^{R}(\theta, p)\, x_4^{S_{24}^{+}} x_5^{S_{25}^{+}}$$

$$A_{11} = \left(\frac{\left(S_{11}^{-}\right)^2}{x_1} + \frac{\left(S_{12}^{-}\right)^2}{x_2} + \frac{\left(S_{13}^{+}\right)^2}{x_3} - (S_1)^2 \right) M_W^{e}$$

$$A_{12} = A_{21} = \left(\frac{\left(S_{11}^{-}\right)\left(S_{21}^{-}\right)}{x_1} + \frac{\left(S_{12}^{-}\right)\left(S_{22}^{-}\right)}{x_2} + \frac{\left(S_{13}^{+}\right)\left(S_{23}^{+}\right)}{x_3} - (S_1)(S_1) \right) M_W^{e}$$

$$A_{122} = \left(\frac{\left(S_{21}^{-}\right)^2}{x_1} + \frac{\left(S_{22}^{-}\right)^2}{x_2} + \frac{\left(S_{23}^{-}\right)^2}{x_3} + \frac{\left(S_{24}^{+}\right)^2}{x_4} + \frac{\left(S_{25}^{+}\right)^2}{x_5} - (S_1)^2 \right) M_W^{e}$$

$$\begin{pmatrix} \dot{\zeta}_1^{a} \\ \dot{\zeta}_2^{a} \end{pmatrix}^{\bullet} = -\begin{pmatrix} J_{11} & J_{12} \\ J_{21} & J_{22} \end{pmatrix} \begin{pmatrix} \dot{\zeta}_1^{a} \\ \dot{\zeta}_2^{a} \end{pmatrix} \qquad J_{11} = \overset{+}{\zeta}_1^{e} A_{11} \qquad J_{12} = J_{21} = \left(\overset{+}{\zeta}_1^{e} \overset{+}{\zeta}_2^{e} \right)^{0.5} A_{12}$$

$$J_{22} = \overset{+}{\zeta}_2^{e} A_{22}$$

Solving $\det\begin{pmatrix} J_{11}-\lambda & J_{12} \\ J_{21} & J_{22}-\lambda \end{pmatrix}=0$ obtains $\lambda_1, \lambda_2 = 0.5\left\{\mathrm{tr}J \pm \left[(\mathrm{tr}J)^2 - 4\det.\right.\right.$

Also, $\lambda_1 = \dfrac{1}{\tau_1}$ $\quad \lambda_2 = \dfrac{1}{\tau_2}$ $\quad \mathrm{tr}J = \lambda_1 + \lambda_2$ $\quad \det J = \lambda_1 \lambda_2$

For $\dfrac{4\det J}{(\mathrm{tr}J)^2} \prec\prec 1$ $\quad \dfrac{1}{\tau_1} \approx -\overset{+e}{\zeta_1} A_{11} - \overset{+e}{\zeta_2} A_{22}$ $\quad \dfrac{1}{\tau_2} \approx -\dfrac{\det J}{\mathrm{tr}J}$

In transformed reaction coordinates,

$$\begin{pmatrix} \dot{\zeta}_1^a \\ \dot{\zeta}_2^a \end{pmatrix} = \begin{pmatrix} M_{11} & M_{12} \\ M_{21} & M_{22} \end{pmatrix} \begin{pmatrix} Z_1 \\ Z_2 \end{pmatrix} \qquad \frac{M_{11}}{M_{21}} = \frac{J_{12}}{\lambda_1 - J_{11}} \qquad \frac{M_{12}}{M_{22}} = \frac{J_{12}}{\lambda_2 - J_{11}} = \frac{\lambda_2 - J_{22}}{J_{21}}$$

Normalizing, $M_{11}^2 + M_{21}^2 = 1$ $\qquad M_{12}^2 + M_{22}^2 = 1$

$$M = \begin{pmatrix} \left(\dfrac{J_{12}^2}{(\lambda_1 - J_{11}) + J_{12}^2}\right)^{0.5} & \left(\dfrac{J_{12}^2}{(\lambda_2 - J_{11}) + J_{12}^2}\right)^{0.5} \\ \left(\dfrac{(\lambda_1 - J_{11})^2}{(\lambda_1 - J_{11}) + J_{12}}\right)^{0.5} & \left(\dfrac{(\lambda_2 - J_{11})^2}{(\lambda_2 - J_{11}) + J_{12}}\right)^{0.5} \end{pmatrix}$$

References

Einstein, A., Sitzber. Deut. Akad. Wiss., Berlin, *Math.-Phys. Kl.* (1929).

Eigen, M., and L. DeMaeyer. "Relaxation Methods." *Technique of Organic Chemistry.* Edited by S. L. Fries, E. S. Lewis, and A. Weissberger, 2nd ed, vol. 8/2. New York: Interscience (1963): 895-1054.

Hertzfeld, K. and T. Litovitz. *Absorption and Dispersion of Ultrasonic Waves*. New York: Academic Press, 1959.

Bernasconi, C. F. *Relaxation Kinetics*. New York: Academic Press, 1976.

Markham, J. J., R. T. Beyer, and R. B. Lindsay. *Reviews of Modern Physics* 23 no. 4 (1951): 353.

G. Schwarz. *Reviews of Modern Physics* 40 (1968): 206.

W. P. Mason,ed. *Physical Acoustics*. vol. 2-A. New York: Academic Press, 1965.

Prigogine, I., and R. Defay. *Chemical Thermodynamics*. Translated by D. H. Everett. London: Longman, 1954.

Sellers, H. S., T. Margulies, and W. H. Schwarz. *Molecular Crystals and Liquid Crystals* 166, no.1 (1988).

Bowen, R. M. *Archive for Rational Mechechanics and Analysis* 24, no. 370 (1967).

—. *Journal of Chemical Physics* 49, no. 1625 (1968).

Mazo, R.M. *Journal of Chemical Physics 28*, no. 1225 (1958).

Margulies, T., and W. Schwarz. *Journal of Chemical Physics 77*, no. 2 (July 15, 1982): 1005.

Hammes, G. G., and P. R. Schimmel. *Enzymes* 2, no.67 (1970).

Castellan, G. W. *Berichte der Bunsen-Gesellschaft Physical Chemistry* 67, no. 898 (1963).

Truesdell, C. A. *Rational Thermodynamics*. New York: Springer, 1984.

Verdier, C., and M. Piau. *JASA* 101, no. 4 (1997).

Coleman, B. D., and W. Noll. "Foundations of Linear Elasticity." *Reviews of Modern Physics 33*, no. 239 (1961).

Debnath, L. *International Journal of Mathematics and Mathematical Sciences* 2003, no. 54 (2003): 3413.

Truesdell, C. A. *Journal of Rational Mechanics and Analysis* 2 (October 1953): 643-741.

Margulies T., and W. Schwarz. *Journal of Acoustical Society of America* 78, 605 (1985).

—. *Journal of Acoustical Society of America* 82, *no. 2* (1987): 522.

Garcia-Colin, L. S., and S. M. T. De La Selva. *Physica* 75, no. 37 (1974).

Bass, H. E., and J. Tan. *JASA* 93, no.1 (1993).

Margulies, T. in *Advances in Nonlinear Acoustics*. Edited by H. Hobaek. Singapore: World Scientific, 1993.

Bagley, R. L. *Journal of Rheology* 27, no. 3 (1983): 201.

Bagley, R. L., and P. J. Torvik. *Journal of Rheology* 30, no.1 (1986): 133.

Ochmann M., and S. Makarov. *JASA* 94, no. 6 (1993).

Oldham, K., and J. Spanier. *The Fractional Calculus*. New York: Academic Press, 2002.

Mainardi, F. *Wave Propagation in Viscoelastic Media*. Boston: Pitman, 1982.

VII

Stochastic Brownian Motions: Statistical Hydrodynamic Analysis of Particle Motions with Fractional Calculus Visco-Elastic Damping Excited by Random Vibrations

Abstract

The purpose of this brief report is to set up and to solve the problem of a randomly vibrating particle with fluid dynamic drag represented with fractional derivatives of the motion variables. This adapts the classical problem of Brownian motion for generalization to and for a better understanding of material drag responses of particle motions in visco-elastics with a fractional stress response model that are subjected to random forcing that previously has not been presented.

Let the dynamical system for differential operator $L_{(t)}$ be represented by $L_{(t)}u(x) = f(t)$ where f(t) denotes time-dependent forcing. The system response for an impulse is obtained by solving $Lu_I(x) = \delta(t)$. A spring, dashpot point-mass vibrating body was previously analyzed where $L_{(t)} \equiv m\frac{\partial^2}{\partial t^2} + 2\varsigma d_{(t)}^{\frac{3}{2}} + k$, and $\varsigma = A\sqrt{\mu\rho}$. The fractional order derivative in the time domain is written $d_{(t)}^{\alpha}(u(x,t) \equiv \frac{1}{\Gamma(1-\alpha)}\frac{d}{dt}\int_0^t \frac{u(x,\tau)}{(t-\tau)^{\alpha}}d\tau$. The $\alpha = \frac{3}{2}$ order differential corresponds to the equation given for displacement as the dependent variable, whereas $\alpha = \frac{1}{2}$ would be for an equivalent equation in terms of velocity, the time derivative of displacement. The solution for the general time-dependent forcing f(t) is obtained by Duhammel's superposition integral, $u(t) = \int_{-\infty}^{\infty} u_I(t-\tau) f(\tau) d\tau$. In Fourier transform space, $U(\omega) = \frac{1}{2\pi}\int_{-\infty}^{+\infty} e^{i\omega t} u(t) dt = U_I(\omega) F(\omega)$. $U_I(\omega)$ is the Fourier transform of $u_I(t)$ multiplied by 2 π.

Now consider a random forcing, that is, the input is a random function, which is stationary with zero mean. The dynamical system equation obtains $L_{(\tau)} u(x) = f(\tau)$, for $L_{(\tau)} \equiv m\frac{\partial^2}{\partial \tau^2} + 2\varsigma D_{(\tau)}^{\frac{3}{2}} + k$. Multiplying this equation by u(t) and then taking the ensemble average yields $L_{(\tau)}\left[\langle u(t)u(t+\tau)\rangle\right] = \langle u(t) f(t+\tau)\rangle$ since the averaging is independent of the time variable t. It is convenient to rewrite this last equation in statistical terminology of auto—and cross covariances. $L_{(\tau)}\left[R_{uu}(\tau)\right] = R_{ff}(\tau)$, $R_{uu}(\tau) = \int_{-\infty}^{\infty} u_I(\tau - t) R_{uf}(t) dt$ $R_{fu}(\tau) = \int_{-\infty}^{\infty} u_I(\tau - t) R_{ff}(t) dt$ referred to as the Weiner-Khinchine relations. Fourier transforming these equations to derive spectra and cross spectra of the input forcing and output system motions give simply,

$$S_{uu}(\omega) = U_I(\omega) S_{uf}(\omega) \quad S_{fu}(\omega) = U_I(\omega) S_{ff}(\omega) \quad S_{uf}(\omega) = \bar{S}_{fu}(\omega) \quad S_{ff}(\omega) = \bar{S}_{ff}(\omega)$$

where the overbar denotes complex conjugate. A Campbell's theorem may be stated from the above, which gives the output power spectral density as a product of the system's frequency response function and the input's power spectral density function

$$S_{uu}(\omega) = U_I(\omega)\bar{U}_I(\omega) S_{ff}(\omega) = |U_I(\omega)|^2 S_{ff}(\omega) \cdot$$

The particle motions for Brownian movements are described by a fractional calculus form; Laplace transformed L{ } and solved using a partial fraction expansion as follows:

$$\frac{dy}{dt}+\beta\frac{d^{\frac{1}{2}}}{dt^{\frac{1}{2}}}y = f(t)$$

$$L\left\{\frac{dy}{dt}+\beta\frac{d^{\frac{1}{2}}}{dt^{\frac{1}{2}}}y\right\} = L\{f(t)\} \qquad L\{y(t)\} = \int_0^\infty e^{-st}y(t)dt \equiv \tilde{y}(s)$$

$$L\left\{\frac{dy}{dt}\right\}+\beta L\left\{\frac{d^{\frac{1}{2}}}{dt^{\frac{1}{2}}}y\right\} = \tilde{f}(s)$$

Transforming and applying the initial condition yields,

$$s\,\tilde{y}(s) - y(0) + \beta s^{\frac{1}{2}}\,\tilde{y}(s) = \tilde{f}(s) \qquad y(0) = 0$$

$$s\,\tilde{y}(s) + \beta s^{\frac{1}{2}}\,\tilde{y}(s) = \tilde{f}(s)$$

Solving for y(s) and finding the inverse transform is summarized as,

$$\tilde{y}(s) = \frac{\tilde{f}(s)}{\left(s+\beta s^{\frac{1}{2}}\right)} = \tilde{f}(s)\left[\frac{1}{\beta}\frac{1}{\sqrt{s}} - \frac{1}{\beta}\frac{\sqrt{s}}{\left(s-\beta^2\right)} + \frac{1}{\left(s-\beta^2\right)}\right]$$

$$y(t) = \int_0^t h_1(t)\,f(t-\tau)\,d\tau \qquad h_1(t) = \exp(\beta^2 t)\,\mathrm{erfc}\left(\beta\sqrt{t}\right.$$

$$\mathrm{erfc}(x) = 1-\mathrm{erf}(x) = 1-\frac{2}{\sqrt{\pi}}\int_0^x e^{-r^2}dr$$

Note the useful expressions:

$$S_y(\omega) = S_f\left|H(j\omega)\right|^2 \qquad H(j\omega) = \frac{1}{(j\omega)+\beta(j\omega)^{0.5}} \qquad j^{0.5} = \frac{\pm(1+j)}{\sqrt{2}}$$

For illustration of these formulas, consider a cosine wave of random amplitude and phase, f(t)= A cos (ω t + φ). The expected value E{f(t)} = E{A} E{cos (ω t + φ)}. For the probability density of the random phase being uniformly distributed between 0 and 2π, the expected value obtains zero. The autocorrelation function

$$R_f(t) = E\{f(t_1)_2\, f(t)\} = E\{A\cos(\omega t_1 + \phi) A\cos(\omega t_2 + \phi)\} =$$

$$= 0.5(\sigma_A^2 + m_A^2)\left\{\cos\left[\omega_0(t_2 - t_1)\right]\right\} + \int_0^{2\pi} \cos\left[\omega_0(t_2 + t_1) + 2u\right] f_\phi(u)du$$

For u uniform phase between 0 and 2 π:

$$R_f(\tau) = E\{f(t) f(t+\tau)\} = 0.5\left(\sigma_A^2 + m_A^2\right)\cos(\omega_0 \tau)$$

For amplitude A of mean value zero,
$R_f(\tau) = 0.5\sigma_A^2 \cos(\omega_0\tau) = R_f(0)\cos(\omega_0\tau)$ and the power spectrum of this forcing signal function becomes,

$$S_x(\omega) = \int_{-\infty}^{\infty} R_f(0)\cos(\omega_0\tau) e^{-j\omega\tau} d\tau = R_f(0)\pi\left[\delta(\omega - \omega_0) + \delta(\omega + \omega_0)\right].$$

In conclusion, the problem of a randomly vibrating particle with fluid drag has been presented and solved using fractional calculus drag. Fractional calculus has been shown to be a fruitful tool for polymer solution rheology representation. Furthermore, showing its simple use in this fundamental problem may help stimulate theoretical and experimental research aimed at understanding turbulent diffusion in complex materials including suspensions and colloids. General classes of material responses may be simply accommodated in this fractional calculus model.

Some useful background information, definitions, and reminders to help in the discussion of the research presented follows. For the output of the j-th experiment $u^{(j)}(t)$ with independent time variable t, the ensemble average denoted by $\langle u(t)\rangle$ is defined by a limit process as follows: $\langle u(t)\rangle \equiv \mathrm{limit}_{N\to\infty} \frac{1}{N}\sum_{j=1}^{N} u^{(j)}(t)$. Also, the ensemble average of a function of u(t), say h(u(t)), may be defined $\langle h(u(t))\rangle \equiv \lim_{N\to\infty} \frac{1}{N}\sum_{j=1}^{N} h(u^{(j)}(t))$.

The variance R_{uu} is defined by forming the moment of realizations of u at two different times t_1 and t_2, $R_{uu}(t_1, t_2) = \langle u(t_1)u(t_2)\rangle$. Time covariances of stationary random functions are even functions that depend only on the time differences $\tau = t_1 - t_2$, that is, the measurements are independent of a translation in time origin. The time origin choice is arbitrary (i.e., the data analysis results will not depend on when the clock starts in recording the data).

Joint or cross covariances of two random functions u(t) and v(t) are defined by $R_{uv}(t) = \langle u(t)v(t+\tau)\rangle$. This is written for a stationary process where $R_{uv}(\tau) = R_{vu}(-\tau)$. Spectra and cross spectra are defined using Fourier transforms. The covariance and cross covariance functions are considered square integrable and approach zero as the time delay increases to large enough values. The power spectral density is $S_{uu}(\omega) = \frac{1}{2\pi}\int_{-\infty}^{\infty} e^{i\omega\tau} R_{uu}(\tau)d\tau$ with $R_{uu}(\omega) = \frac{1}{2\pi}\int_{-\infty}^{\infty} e^{-i\omega\tau} S_{uu}(\tau)d\tau$.

References

Bagley, R. L., and P. J. Torvik. *Shock and Vibration Bulletin* 49, part 2 (September 1979): 135-143.

Torvik, P. J., and R. L. Bagley. *Shock and Vibration Bulletin*. part 2 (1985): 81-84.

O'Neil, P. V. *Advanced Engineering Mathematics*. California: Wadsworth Inc., 1983.

Bendat, J. S., and A. G. Piersol. *Random Data and Analysis*. New York: Wiley-Interscience, 2000.

Carl W. Helstrom. *Probability and Stochastic Processes for Engineers*. New York: Macmillan, 1984.

Haddad, Abraham. *Probabilistic Systems and Random Signals*. New Jersey: Prentice Hall, 2006.

Landahl, M. T., and E. Mollo-Christensen. *Turbulence and Random Processes in Fluid Mechanics*. Cambridge: Cambridge University Press, 1986.

Einstein, A. *Investigations in the Theory of Brownian Motion*. R. Firth and A. D. Cooper translation. New York: Dover, 1956.

Perrin, J. B. *Brownian Movement and Molecular Reality*. Translated by F. Soddy. London: Taylor & Francis, 1910.

Csanady, C. S. *Turbulent Diffusion in the Environment*. Holland: D. Reidel Publishing, 1973.

VIII

Multicomponent Electrophoretic Transport with Memory and Chemical Reactions

Abstract:

Fick's law of diffusion is generalized to include the rate of gradient changes similar to a Cattaneo memory in heat conduction. First, for a non-reacting binary mixture, the new Fick's law includes this additional inertial term. A one-dimensional diffusion problem is presented and solved using a separation of variables technique with that term. Then tertiary and quintic component systems are addressed including illustrative reaction kinetics. Interpreting the final results gives another contribution to the role of non-ideal activity from diffusive modes that contain the new inertial diffusive term in Fick's law.

I. A Non-reacting Diffusive System

Let the mass flux be expressed as follows: $J_i = -D\left(\frac{\partial c}{\partial x_i} - \tau\left(\frac{\partial c}{\partial x_i}\right)^{\bullet}\right)$. Then

$J_i = -D\left[1-\tau\frac{\partial}{\partial t}\right]\left(\frac{\partial c}{\partial x_i}\right)$ or $\frac{J_i}{\left[1-\tau\frac{\partial}{\partial t}\right]} = -D\left(\frac{\partial c}{\partial x_i}\right)$. Since $\frac{1}{1-x} \approx 1+x$,

$\left[1+\tau\frac{\partial}{\partial t}\right]J_i = -D\left(\frac{\partial c}{\partial x_i}\right)$. However remembering the general balance law,

$\frac{\partial c}{\partial t}+\frac{\partial J_i}{\partial x_i}=0$ in one dimension includes an inertial term with coefficient τ. This obtains the following dynamical equation of motion for the component concentration c and diffusion coefficient D:

$$\tau\frac{\partial^2 c}{\partial t^2}+\frac{\partial c}{\partial t}=D\frac{\partial^2 c}{\partial x^2} \qquad c(x=0,t)=0; \qquad c(x=L,t)=0; \qquad c(x,t=0)=c_0;\ \left.\frac{\partial c}{\partial t}\right|_{t=0}=0$$

With the given boundary and initial conditions and using a separation of time and space functions $c(x,t)=X(x)T(t)$, the above equation can be solved. Rewriting after diving by DXT:

$\frac{\tau}{D}\frac{T''}{T}+\frac{1}{D}\frac{T'}{T}=\frac{X''}{X}$. Set both sides equal to $-\lambda$ where the constant $\lambda>0$.

Examining the temporal equation, $T''+\frac{1}{\tau}T'=\frac{-\lambda}{\tau}DT$ and solving for exponential solutions yield the equation

$$T=e^{mt}; \qquad m^2+\frac{1}{\tau}m+\frac{\lambda D}{\tau}=0$$

$$m_{1,2}=\frac{1}{2\tau}(-1\pm\sqrt{1-4D\lambda\tau})$$

For constants A and B, $T(t)=Ae^{m_1t}+Be^{m_2t}$.

Next, the spatial function satisfies $\frac{d^2X}{dX^2}=-\lambda X$.

$$X(x)=A_1\sin\left(\sqrt{\lambda}x\right)+A_2\cos\left(\sqrt{\lambda}x\right).$$

The expression c(x=0,t) = 0 implies that $A_2=0$ while c(x=L,t) = 0 restricts

λ such that $\sqrt{\lambda}L=n\pi \quad \sqrt{\lambda}=\frac{n\pi}{L}\equiv\lambda_n \quad n=1,2,3,\ldots$ Then

$$c(x,t)=\sum W_n\sin\left(\frac{n\pi}{L}x\right)\left(e^{m_1t}-\beta e^{m_2t}\right); \quad \beta\equiv\frac{m_1}{m_2}; \quad W_n=aA_n \ .$$

As in a Fourier series expansion, the coefficients can be obtained from,

$$W_n = \frac{1}{L}\int_{-L}^{L} c(x)\sin\left(\frac{n\pi}{L}x\right)dx = 4\frac{c_0}{n\pi};\ \text{n odd} \quad ;\ \text{Let } n = 2m+1,\ m=0, 1, 2, 3\ldots$$

$$c(x,t) = \sum_{m=1} \frac{4c_0}{m\pi}\sin\left(\frac{m\pi}{L}x\right)\left(e^{m_1 t} - \beta e^{m_2 t}\right);\quad m_{1,2} = \frac{1}{2\tau}(-1 \pm \sqrt{1-4D\lambda_m\tau})$$

II. A Ternary Diffusive System

In vector matrix notation, the diffusion of a ternary system (3 species) with new inertial terms previously motivated may be written,

$$\begin{pmatrix} \tau_1 & 0 \\ 0 & \tau_2 \end{pmatrix}\frac{\partial^2}{\partial t^2}\begin{pmatrix} c_1 \\ c_2 \end{pmatrix} + \frac{\partial}{\partial t}\begin{pmatrix} c_1 \\ c_2 \end{pmatrix} = \begin{pmatrix} D_{11} & D_{12} \\ D_{21} & D_{22} \end{pmatrix}\nabla^2\begin{pmatrix} c_1 \\ c_2 \end{pmatrix}$$

or in compact notation, $\underline{\underline{T}}\vec{c}'' + \vec{c}' = \underline{\underline{D}}\nabla^2\vec{c}$. In steady state, a Laplace equation is obtained for each component, that is, $\nabla^2\vec{c} = 0 = \nabla^2 c_i$.

The time-dependent equation may be separated into independent equations by a transformation that diagonalizes the diffusion matrix as follows. Let

$$\tilde{c} = \underline{\underline{Q}}^{-1}\vec{c},\ \underline{\underline{Q}}^{-1}\underline{\underline{D}}\,\underline{\underline{Q}} = \underline{\underline{\Lambda}} \equiv \begin{pmatrix} \lambda_1 & 0 \\ 0 & \lambda_2 \end{pmatrix}; \qquad \underline{\underline{Q}} = \begin{pmatrix} \dfrac{D_{11} - D_{22} - \Delta}{2D_{21}} & \dfrac{D_{11} - D_{22} + \Delta}{2D_{21}} \\ 1 & 1 \end{pmatrix}$$

$$\underline{\underline{Q}}^{-1} = \begin{pmatrix} -\dfrac{D_{21}}{\Delta} & \dfrac{D_{11} - D_{22} + \Delta}{2\Delta} \\ \dfrac{D_{21}}{\Delta} & -\left(\dfrac{D_{11} - D_{22} - \Delta}{2\Delta}\right) \end{pmatrix}; \qquad \Delta = \sqrt{(\mathrm{tr}\,\underline{\underline{D}})^2 - 4\det\underline{\underline{D}}} = \sqrt{(D_{11} - D_{22})^2 + 4D_{12}D_{21}}$$

$$\underline{\underline{T}} = \begin{pmatrix} \tau_1 & 0 \\ 0 & \tau_2 \end{pmatrix}; \qquad \underline{\underline{Q}}^{-1}\underline{\underline{T}} = \underline{\underline{T}}\,\underline{\underline{Q}}^{-1}; \qquad \text{then} \quad \underline{\underline{T}}\tilde{c}'' + \tilde{c}' = \underline{\underline{\Lambda}}\nabla^2\tilde{c}$$

For chemical reactions, a matrix $\underline{\underline{K}}$ representing rate functions (with possible equilibrium concentrations) appears in the above equation. When the equations are linearized about equilibrium, $c_i^0 \quad c_i = c_i^0 + \delta_i$ a simultaneous diagonalization of matrices may be implemented to separate the equations.

$\underline{\underline{T}}\delta\tilde{c}'' + \delta\tilde{c}' + \underline{\underline{\Delta}}\nabla^2\delta\tilde{c} + \underline{\underline{K}}\delta\tilde{c}$ $\left|\underline{\underline{K}} - \lambda\underline{\underline{\Delta}}\right| = 0$ Solving for the eigenvalues λ_i enables linearly independent eigenvectors $\underline{x}_i$ to be formed to reduce simultaneously the matrices to diagonal form, for $\underline{\Delta}$ is positive definite and both matrices $\underline{\Delta}$ and $\underline{K}$ are Hermitian (symmetric in real case). $\underline{\underline{K}}\,\underline{x}_i = \lambda_i \Lambda \underline{x}_i$ $\quad \underline{\underline{R}} \equiv [\underline{x}_1, \underline{x}_2, \underline{x}_3 \ldots, \underline{x}_n]$ $\quad \underline{\underline{R}}^H \underline{\underline{K}}\,\underline{\underline{R}}$ and $\underline{\underline{R}}^H \underline{\underline{\Delta}}\underline{\underline{R}}$ are diagonal matrices, which may be normalized.

III. A Ternary Reacting Diffusing System

Consider the chemical reaction comprising ion-pair formation and decomposition:

$$A^{z-} + B^{z+} \leftrightarrow \left[C^{z+}A^{z-}\right]^0 \quad \text{or} \quad Y_1 + Y_2 \leftrightarrow Y_3 \quad Y_3 = [Y_1 Y_2]^0$$

The balances of mass with reaction kinetics and diffusion including an inertial diffusive term follows. The variables c_j represent the concentrations in units of molar density [mol/m^3]. The effect of an inertial diffusive term forms a generalization of Fick's law for the spreading diffusive motion. Other phoretic contributions such as by temperature and pressure are considered negligible.

$$\tau_1 \frac{\partial^2 c_1}{\partial t^2} + \frac{\partial c_1}{\partial t} + \nabla \bullet J_1 = -k_{12} c_1 c_2 + k_3 c_3$$

$$\tau_2 \frac{\partial^2 c_2}{\partial t^2} + \frac{\partial c_2}{\partial t} + \nabla \bullet J_2 = -k_{12} c_1 c_2 + k_3 c_3$$

$$\tau_3 \frac{\partial^2 c_3}{\partial t^2} + \frac{\partial c_3}{\partial t} + \nabla \bullet J_3 = k_{12} c_1 c_2 - k_3 c_3$$

In equilibrium, $k_{12} c_1^0 c_2^0 = k_3 c_3^0$.

The flux terms for electrophoretic and concentration gradient motions become for an electric field strength $\underline{E}$:

$$J_i = -D_i \nabla c_i + c_i z_i e \frac{D_i}{kT} \underline{E} + \kappa_1 \nabla \theta + \kappa_2 \nabla p$$

$$\nabla \cdot J_i = -D_i \nabla^2 c_i + z_i e \frac{D_i}{kT} \left(c_i \nabla \cdot E + \underline{E} \nabla \cdot c_i \right) + \kappa_{1i} \nabla^2 \theta + \kappa_{2i} \nabla^2 p$$

$$\nabla \cdot \underline{E} = \frac{N_A e}{\varepsilon_0 \varepsilon} \sum_{i=1,2} z_i \delta c_i = \frac{N_A e}{\varepsilon_0 \varepsilon} \left(z_1 \delta c_1 + z_2 \delta c_2 \right)$$

$$\nabla \cdot \delta J_1 = -D_1 \nabla^2 \delta c_1 + q_1^2 D_1 \delta c_1 + q_1^2 \frac{z_2}{z_1} D_1 \delta c_2 + \kappa_{11} \nabla^2 \theta^a + \kappa_{21} \nabla^2 p^a$$

$$\nabla \cdot \delta J_2 = -D_2 \nabla^2 \delta c_2 + q_2^2 D_2 \delta c_2 + q_2^2 \frac{z_1}{z_2} D_2 \delta c_1 + \kappa_{12} \nabla^2 \theta^a + \kappa_{22} \nabla^2 p^a$$

$$\nabla \cdot \delta J_3 = -D_3 \nabla^2 \delta c_3 + \kappa_{1i} \nabla^2 \theta^a + \kappa_{23} \nabla^2 p^a \qquad z_3 = 0$$

$$q_1^2 \equiv \frac{e^2 z_1^2 N_A}{\varepsilon_0 \varepsilon kT} c_1^0, \qquad q_2^2 \equiv \frac{e^2 z_2^2 N_A}{\varepsilon_0 \varepsilon kT} c_2^0, \qquad q_1^2 + q_2^2 = \chi^2$$

The first-order equations are derived about equilibrium values of c_i^0 , $c_i = c_i^0 + c_i^a$ to obtain,

$$\left(\tau_1 \frac{\partial^2}{\partial t^2} + \left[\frac{\partial}{\partial t} \right] - D_1 \nabla^2 + q_1^2 D_1 + k_{12} c_2^0 \right) c_1^a + \left(q_1^2 D_1 \frac{z_2}{z_1} + k_{12} c_1^0 \right) c_2^a - k_3 c_3^a - \kappa_{11} \nabla^2 \theta^a - \kappa_{21} \nabla^2 p^a = 0$$

$$\left(\tau_2 \frac{\partial^2}{\partial t^2} + \left[\frac{\partial}{\partial t} \right] - D_2 \nabla^2 + q_2^2 D_2 + k_{12} c_1^0 \right) c_2^a + \left(q_2^2 D_2 \frac{z_1}{z_2} + k_{12} c_2^0 \right) \delta c_1 - k_3 c_3^a - \kappa_{12} \nabla^2 \theta^a - \kappa_{22} \nabla^2 p^a = 0$$

$$-k_{12} c_2^0 c_1^a - k_{12} c_1^0 c_2^a + \left(\tau_3 \frac{\partial^2}{\partial t^2} + \frac{\partial}{\partial t} - D_3 \nabla^2 + k_3 \right) c_3^a - \kappa_{13} \nabla^2 \theta^a - \kappa_{23} \nabla^2 p^a = 0$$

Here applying Fourier-Laplace transforms,

$$c(\vec{q}, t) = \int c(\vec{x}, t) e^{(i \vec{q} \bullet \vec{x})} d\vec{x}; \qquad c(\vec{q}, s) = \int_0^\infty c(\vec{q}, t) e^{-st} dt$$

obtains a matrix equation with algebraic elements.

$$\underline{C} \equiv \begin{pmatrix} c_1^a \\ c_2^a \\ c_3^a \\ \theta^a \\ p^a \end{pmatrix} \qquad \underline{\underline{A}}\,\underline{C} = \underline{C}_0 \qquad \underline{C}_0 \equiv \begin{pmatrix} \tau_1 s\delta c_1 + \tau_1 \delta c_1' \\ \tau_2 s\delta c_2 + \tau_2 \delta c_2' \\ \tau_3 s\delta c_3 + \tau_3 \delta c_3' \end{pmatrix}$$

$$\underline{\underline{A}} = \begin{pmatrix} \tau_1^2 s^2 + s + \left(q_1^2 + q^2\right) D_1 + k_{12} c_2^0 & q_2^2 D_{11} \frac{z_2}{z_1} + k_{12} c_1^0 & -k_{13} & -\kappa_{11} & -\kappa_{21} \\ q_2^2 D_2 \frac{z_1}{z_2} + k_{12} c_2^0 & \tau_2 s^2 + s + \left(q_2^2 + q^2\right) D_2 + k_{12} c_1^0 & -k_{13} & -\kappa_{12} & -\kappa_{22} \\ -k_{12} c_2^0 & -k_{12} c_1^0 & \tau_3 s^2 + s + q^2 D_3 + k_3 & -\kappa_{13} & -\kappa_{23} \\ & & & & \end{pmatrix}$$

The determinant equation det $\underline{\underline{A}} = 0$ results in a polynomial equation in s. The diffusional modes are typically found by setting $s = -Dq^2$ and ignoring higher terms.

Consider $\kappa_{ij} = 0$ for a symmetrical electrolyte, $z_1 = -z_2$,

$$D = \frac{k_3 D_1 D_2 \left(q_1^2 + q_2^2\right) + k_{12} D_3 \left(c_1^0 + c_2^0\right)\left(q_1^2 D_1 + q_2^2 D_2\right)}{\left(k_3 + k_{12} c_1^0 + k_{12} c_2^0\right)\left(q_1^2 D_1 + q_2^2 D_2\right)}$$

This can be interpreted as the sum of the diffusion coefficients of the free and associated ions.

$$D = \frac{D_F}{1 + \frac{k_{12}\left(c_1^0 + c_2^0\right)}{k_3}} + \frac{D_A}{1 + \frac{k_3}{k_{12}\left(c_1^0 + c_2^0\right)}}; \qquad D_A = \frac{D_1 D_2 \left(q_1^2 + q_2^2\right)}{q_1^2 D_1 + q_2^2 D_2}; \qquad D_A = D_3$$

It is usually assumed that the non-ideal activity contributes to higher order where both the diffusional coefficients would be modified, for example, $D = D_A^0 \left(1 + \frac{\partial \ln \gamma}{\partial \ln c}\right)$.

Therefore, the new continuum inertial diffusive term would contribute to this non-ideality.

IV. A Quintic-Reacting Diffusing System

Consider the diffusional transport of unsymmetrical electrolytes in solution. Let a 2-1 electrolyte or polyelectrolyte, solvation-desolvation process, and aggregation or decomposition states exist as represented by the simultaneous reactions:

$$C^{z+} + A^{z-} \Leftrightarrow C^{z+}(S_2)A^{z-} \Leftrightarrow C^{z+}(S)A^{z-} \Leftrightarrow C^{z+}A^{z-}$$

$$Y_5 + Y_6 \Leftrightarrow Y_2 \Leftrightarrow Y_3 \Leftrightarrow Y_4$$

The linearized diffusion-reaction equations with the inertial diffusive terms obtain,

$$\tau_2 \frac{\partial^2 \delta c_2}{\partial t^2} + \frac{\partial \delta c_2}{\partial t} + \nabla \bullet \delta J_2 = k_{52}\delta c_5 + k_{62}\delta c_6 - k_{21}\delta c_2 - k_{23}\delta c_2 + k_{32}\delta c_3$$

$$\tau_3 \frac{\partial^2 \delta c_3}{\partial t^2} + \frac{\partial \delta c_3}{\partial t} + \nabla \bullet \delta J_3 = k_{23}\delta c_2 - k_{32}\delta c_3 - k_{34}\delta c_3 + k_{43}\delta c_4$$

$$\tau_4 \frac{\partial^2 \delta c_4}{\partial t^2} + \frac{\partial \delta c_4}{\partial t} + \nabla \bullet \delta J_4 = k_{34}\delta c_3 - k_{43}\delta c_4$$

$$\tau_5 \frac{\partial^2 \delta c_5}{\partial t^2} + \frac{\partial \delta c_5}{\partial t} + \nabla \bullet \delta J_5 = -k_{52}\delta c_5 - k_{62}\delta c_6 - k_{12}\delta c_2$$

$$\tau_6 \frac{\partial^2 \delta c_6}{\partial t^2} + \frac{\partial \delta c_6}{\partial t} + \nabla \bullet \delta J_6 = -k_{52}\delta c_5 - k_{62}\delta c_6 + k_{12}\delta c_2$$

with $k_{52} = k_{12}c_6^0$; $\quad k_{62} = k_{12}c_5^0$

$$\nabla\cdot\delta J_2 = -D_2\nabla^2\delta c_2 + q_2^2 D_2\delta c_2 + q_2^2\frac{z_3}{z_2}D_2\delta c_3 + q_2^2\frac{z_4}{z_2}D_2\delta c_4 + q_5^2 D_2\frac{z_5}{z_2}\delta c_5 + q_2^2\frac{z_6}{z_2}D_2\delta c_6$$

$$\nabla\cdot\delta J_3 = q_3^2 D_3\frac{z_2}{z_3}\delta c_2 - D_3\nabla^2\delta c_3 + q^2 D_3\delta c_3 + q_3^2\frac{z_4}{z_3}D_3\delta c_4 - q_3^2 D_3\frac{z_5}{z_3}\delta c_5 + q_3^2\frac{z_6}{z_3}D_3\delta c_6$$

$$\nabla\cdot\delta J_4 = q_4^2 D_4\frac{z_2}{z_4}\delta c_2 + q_4^2 D_4\frac{z_3}{z_4}\delta c_3 - D_4\nabla^2\delta c_4 + q_4^2 D_4\delta c_4 + q_4^2 D_4\frac{z_5}{z_4}\delta c_5 + q_4^2\frac{z_6}{z_4}D_4\delta c_6$$

$$\nabla\cdot\delta J_5 = q_5^2 D_5\frac{z_2}{z_5}\delta c_2 + q_5^2 D_5\frac{z_3}{z_5}\delta c_3 + q_5^2\frac{z_4}{z_5}D_5\delta c_4 - D_5\nabla^2\delta c_5 + q_5^2 D_5\delta c_5 + q_5^2\frac{z_6}{z_5}D_5\delta c_6$$

$$\nabla\cdot\delta J_6 = q_6^2 D_6\frac{z_2}{z_6}\delta c_2 + q_6^2 D_6\frac{z_3}{z_6}\delta c_3 + q_6^2\frac{z_4}{z_6}D_3\delta c_4 + q_6^2 D_6\frac{z_5}{z_6}\delta c_5 - D_6\nabla^2\delta c_6 + q_6^2 D_6\delta c_6$$

As before, these partial differential equations can be Fourier and Laplace transformed to yield a set of algebraic equations with the transformed initial conditions to solve this problem with the additional generalized Fick's law term.

V. Discussion and Conclusion

The purpose of this investigation is fulfilled in presenting the inertial motion term in the classical form of Fick's law similar to a heat conduction problem with memory and solving several problems to help motivate further experiments for validation and verification. The neglect of this term, which arises from continuum theory, may find importance for nanotechnology applications, material properties estimation, or possible ultrasound monitoring or light scattering measurement. Furthermore, a diagonalization procedure is introduced to diagonalize simultaneously the diffusion and kinetics matrices that has not previously been identified in this area of application.

Additional insights to diffusive dynamics may also be made when reformulating in terms of fractional calculus. For example, in vector matrix notation, the diffusion of a ternary system (3 species) without inertial terms simplify to,

$$\frac{\partial}{\partial t}\begin{pmatrix} c_1 \\ c_2 \end{pmatrix} = \begin{pmatrix} D_{11} & D_{12} \\ D_{21} & D_{22} \end{pmatrix}\nabla^2\begin{pmatrix} c_1 \\ c_2 \end{pmatrix}$$

or in compact notation with an accent for time differentiation. As before, the time-dependent equation may be separated into independent equations by a transformation that diagonalizes the diffusion matrix as follows: Let

$$\tilde{c} = \underline{\underline{Q}}^{-1}\vec{c}, \qquad \underline{\underline{Q}}^{-1}\underline{\underline{D}}\underline{\underline{Q}} = \underline{\underline{\Lambda}} \equiv \begin{pmatrix} \lambda_1 & 0 \\ 0 & \lambda_2 \end{pmatrix};$$

$$\text{then} \quad \tilde{c}' = \underline{\underline{\Lambda}}\nabla^2\tilde{c}$$

That is, $\frac{\partial \tilde{c}_1}{\partial t} = \lambda_1 \frac{\partial^2 \tilde{c}_1}{\partial x^2}$ and $\frac{\partial \tilde{c}_2}{\partial t} = \lambda_2 \frac{\partial^2 \tilde{c}_2}{\partial x^2}$ or

$$\frac{\partial^2 V_1}{\partial u_1^2} = \frac{\partial V_1}{\partial t} \qquad \frac{\partial^2 V_2}{\partial u_1^2} = \frac{\partial V_2}{\partial t} \qquad u_1 = \frac{x}{\sqrt{\lambda_1}}, \qquad u_2 = \frac{x}{\sqrt{\lambda_2}}$$

$$V_i = \tilde{c}_i(x,t) - \tilde{c}_i^0, \qquad i = 1,2$$

These may also be expressed as $\frac{\partial V_1}{\partial u_1} = -\frac{\partial^{1/2} V_1}{\partial t^{1/2}}$ $\quad \frac{\partial V_2}{\partial u_2} = -\frac{\partial^{1/2} V_2}{\partial t^{1/2}}$

VI. References

Fick, A. *Annalen der Physik* 94, no.59 (1855).

Huang, M-N (1973), Ph.D. Thesis, Johns Hopkins University.

Margulies, T., and W. Schwarz. "A Multiphase Continuum Theory for Sound Wave Propagation through Emulsions and Colloids Using a Generalized Fick's Law of Diffusion." *Journal of Acoustical Society of America* 91, no.6 (1991): 3209.

Mueller, I., and Ruggeri. *Rational Extended Thermodynamics*. Second edition, New York: Springer Verlag, 1998.

O'Neil, P. *Advanced Engineering Mathematics*. California: Wadsworth Inc., 1983.

Noble, B. *Applied Linear Algebra.* New Jersey: Prentice Hall, 1969.

Balluffi, R., S. Allen, and C. Carter. *Kinetics of Materials.* New Jersey: Wiley and Sons, 2005.

Turq, P., J. Barthel, and M. Chemla. "Transport, Relaxation, and Kinetic Processes in Electrolyte Solutions." Lecture notes.

Berne, B., and R. Pecora. *Dynamic Light Scattering.* New York: Dover, 1976.

Haase, R. *Thermodynamics of Irreversible Processes.* Massachusetts: Addison-Wesley, 1969.

De Groot, S., and P. Mazur. *Nonequilibrium Thermodynamics.* Amsterdam: North Holland, 1969.

Tien, C. J., A. Majundar, and Gerner, A. *Microscale Energy Transport.* New York: Taylor & Francis, 1998.

S. Simons, *Am. J. Phys., 54*, 11, November 1986.

Appendix: Brief note on simultaneous diagonalization of two matrices:

For Hermitian matrices $\underline{\underline{A}}, \underline{\underline{B}}$, where $\underline{\underline{B}}$ is positive definite and λ_i are roots i=1,2, . . . n of the polynomial equation, $\det(\underline{\underline{A}} - \lambda \underline{\underline{B}}) = 0$, there exist linearly independent vectors $\underline{x}_i$ such that $\underline{\underline{A}}\underline{x}_i - \lambda_i \underline{\underline{B}}\underline{x}_i = 0,\ i = 1,..n \quad (\underline{x}, \underline{\underline{B}}\underline{x}) = \underline{\underline{I}}$.

Let $R = \left[\underline{x}_1, \underline{x}_2, ...\underline{x}_n\right]$ so then $\underline{\underline{R}}^H \underline{\underline{A}}\underline{\underline{R}} = \underline{\underline{\Lambda}} \qquad \underline{\underline{R}}^H \underline{\underline{B}}\underline{\underline{R}} = \underline{\underline{I}}$.

$\underline{\underline{\Lambda}}$ is a matrix whose diagonal elements are λ_i and $\underline{\underline{I}}$ is the identity matrix with ones along the diagonal and zeros elsewhere. The complex conjugate of the ordinary matrix transpose $\underline{\underline{A}}^H \equiv \overline{\underline{\underline{A}}}^T$ is called a Hermitian transpose. This helps define a Hermitian matrix as such that $\underline{\underline{A}}^H = \underline{\underline{A}}$. The inner product (or scalar product) of two real $n \times 1$ column vectors is a scalar defined by $(\underline{z}, \underline{w}) = z_1 w_1 + z_2 w_2 + z_3 w_3 + .. + z_n w_n$ and the norm or length is defined by, $\|\underline{u}\| = (\underline{u}, \underline{u})^{\frac{1}{2}} = \left(u_1^2 + u_2^2 + u_3^2 + ..u_n\right)^{\frac{1}{2}}$.

IX

Nonlinear Visco-Thermal Acoustic Waves

Using balance equations for total mass, linear momentum, and entropy for an irreversible thermodynamic system, a nonlinear wave equation derivation is summarized below (for the x-direction).The analysis considers a visco-thermal equation with a fractional calculus rheology model and a material with thermal memory. Comparison to earlier investigator's Newtonian viscous, Fourier heat conduction cases are made. The balances of total mass and linear momentum in differential form for position x are,

$$\rho_{,t} + \left(\rho v_{,x}\right) = 0$$

$$\left(\rho v\right)_{,t} = -\left[\rho v^2 + p - S_{xx}\right]_{,x}$$

Next, write the working with the entropy (or energy) equation,

$$\rho_0 \theta_0 \eta^a_{,t} + \rho_0 \theta_0 \eta^a_{,x} = h_{,x}$$

Constitutive assumptions for stress and heat conduction responses are needed to define the system such as,

$$S_{xx} = E_0 \frac{\partial u}{\partial x} + E_1 \frac{\partial^\alpha}{\partial t^\alpha} \frac{\partial u}{\partial x}; \qquad v \equiv \frac{\partial u}{\partial t}$$

for the stress tensor and where the heat flux vector may be written in the form including memory,

$$h = -\frac{k_\theta}{\tau_c} \int e^{\frac{\Delta\theta}{\tau_c}} \theta_{,x} \, d\theta_p$$

This provides a needed expression in this derivation,

$$\eta^a_{,x} = T \left[\int e^{\Delta\theta / \tau_c} \theta_{,x} \, d\theta_p \right]_{,x} \qquad T \equiv \frac{k_\theta p_{,\eta}}{\tau_c c_0 \rho_0 \theta_0}$$

Differentiating the mass and momentum balances with respect to time and space, respectively,

$$(\rho v)_{,xt} = -\left[\rho v^2 + p - S_{xx} \right]_{,xx}$$

$$\rho_{,tt} = -(\rho v)_{,tx} \qquad \rho_{,tt} - \left[\rho v^2 + p - S_{xx} \right]_{,xx} = 0$$

and expanding the pressure function in a Taylor series and then differentiating yields,

$$p^a = c_0^2 \rho^a + 0.5 \left(c_0^2 \right)_{,\rho} \left(\rho^a \right)^2 + p_{,\eta} \eta^a$$

$$p^a_{,x} = c_0^2 \rho^a_{,x} + \left(c_0^2 \right)_{,\rho} \left(\rho^a \right) \rho^a_{,x} + p_{,\eta} \eta^a_{,x}$$

to obtain the following linear momentum equation:

$$(\rho u)_{,xtt} = -\left[\rho u_{,t}^{2} + c_0^{2}\rho^{a} + 0.5\left(c_0^{2}\right)_{,\rho}\left(\rho^{a}\right)^{2} + p_{,\eta}\eta^{a} + E_0\frac{\partial u}{\partial x} + E_1\frac{\partial^{\alpha}}{\partial t^{\alpha}}\frac{\partial u}{\partial x}\right]_{,xx}$$

Subtracting $-c_0^{2}\rho_{,xx}$ from both sides of the equation and helps generate a wave equation,

$$(\rho u)_{,xtt} - c_0^{2}\rho_{,xx} = -\left[c_0^{2}\rho^{a} + \left(\frac{c_0^{2}}{\rho_0} + 0.5\left(c_0^{2}\right)_{,\rho}\right)\left(\rho^{a}\right)^{2} + T\left[\int e^{\Delta\theta/\tau} c_{\theta,x}\, d\theta_{p}\right] + E_0\frac{\partial u}{\partial x} + E_1\frac{\partial^{\alpha}}{\partial t^{\alpha}}\frac{\partial u}{\partial x}\right]_{,xx} - c_0^{2}\rho_{,xx}$$

Integrating this on both sides with respect to x. The constant is set to zero for free waves but would equal the contribution from external sources of radiation.

Define new coordinates $z = x - c_0 t, \quad t = \tau$.

$$\frac{\partial}{\partial z} = \frac{\partial}{\partial x}, \qquad \frac{\partial}{\partial t} = -c_0\frac{\partial}{\partial z} + \frac{\partial}{\partial \tau} \cdot$$

$$\left(\frac{\partial}{\partial t} - c_0\frac{\partial}{\partial x}\right)\left(\frac{\partial}{\partial t} + c_0\frac{\partial}{\partial x}\right) = -2c_0\frac{\partial^{2}\rho^{a}}{\partial z\partial \tau} + \frac{\partial^{2}\rho^{a}}{\partial \tau^{2}} \approx -2c_0\frac{\partial^{2}\rho^{a}}{\partial z\partial \tau}$$

neglecting $\frac{\partial^2 \rho}{\partial \tau^2}$ for a slow wave profile progress, which for a plane progressive wave enables the conversion of derivatives, $\left[g\right]_{,t} \cong -c_0\left[g\right]_{,x}$.

$$\frac{\partial \rho^a}{\partial \tau} = \frac{1}{2c_0}\left[c_0^2 \rho^a + \left(\frac{c_0^2}{\rho_0} + 0.5\left(c_0^2\right)_{,\rho}\right)\left(\rho^a\right)^2 + T\left[\int e^{\Delta\theta/\tau} c_{\theta_{,x}}\, d\theta_p\right] + \right.$$

$$\left. + E_0 \frac{\partial u}{\partial x} + E_1 \frac{\partial^\alpha}{\partial t^\alpha}\frac{\partial u}{\partial x}\right]_{,x} - c_0^2 \rho_{,x}$$

Incorporating the linear acoustic relation $u_{,t} = \frac{c_0}{\rho_0}\rho^a$,

$$\frac{\partial u}{\partial \tau} = \left[\left(1 + 0.5\frac{\rho_0\left(c_0^2\right)_{,\rho}}{c_0^2}\right)uu_{,x} + \frac{T\Omega}{2\rho_0}\left[\int_{-\infty}^{u} e^{\Omega\Delta u_{,t}/\tau} c_{u_{,txx}}\, du'\right] + \right.$$

$$\left. + \frac{E_0}{2\rho_0}\frac{\partial^2 u}{\partial x^2} + \frac{E_1}{2\rho_0}\frac{\partial^\alpha}{\partial t^\alpha}\frac{\partial^2 u}{\partial x^2}\right] \qquad \Omega \equiv \frac{\theta_0 \beta}{\rho_0 c_p}$$

For $\tau_c/\Omega << 1$

$$\frac{\partial u}{\partial \tau} = \left[\left(1 + 0.5\frac{\rho_0\left(c_0^2\right)_{,\rho}}{c_0^2}\right)uu_{,x} + \frac{T\tau_c}{2\rho_0}u_{,txx} + \frac{E_0}{2\rho_0}\frac{\partial^2 u}{\partial x^2} + \frac{E_1}{2\rho_0}\frac{\partial^\alpha}{\partial t^\alpha}\frac{\partial^2 u}{\partial x^2}\right.$$

If $\alpha = 1$, $E_0 = 0$, the classical equation for a visco-thermal nonlinear wave is obtained.

$$\frac{\partial u}{\partial \tau} = \left(1 + 0.5 \frac{\rho_0 \left(c_0^2 \right)_{,\rho}}{c_0^2} \right) uu_{,x} + \left(\frac{T\tau_c}{2\rho_0} + \frac{E_1}{2\rho_0} \right) u_{,txx}$$

For considering chemical reactions, the pressure perturbation has been generalized as follows for a reaction volume change,

$$p^a \approx c_0^2 \rho^a + \beta_1 \rho^{a2} + \frac{\partial p}{\partial \eta} \eta^a + \sum_{i=1,3} \frac{\partial p}{\partial B_i} b_i - \int_{-\infty}^{t} G(t-t') \frac{\partial \rho^a}{\partial t} dt + p_{,\eta} \eta^a$$

$$G(t) = m_1 c_0^2 \exp\left[-(t-t')/\tau_{\mathrm{chem1}} \right] \qquad m_1 \equiv \frac{c_\infty^2 - c_0^2}{c_0^2}$$

For two independent reactions:

$$G(t) = m_1 c_0^2 \exp\left[-(t-t')/\tau_{\mathrm{chem1}} \right] + m_2 c_0^2 \exp\left[-(t-t')/\tau_{\mathrm{chem2}} \right]$$

$$c_\infty^2 = c_0^2 \left(1 + m_1 + m_2 \right)$$

Heats of reaction from a temperature form of the energy equation may be included as well.

References

Naugolnykh, K., and L. Ostrovsky. *Nonlinear Wave Processes in Acoustics*. New York: University of Cambridge, 1998.

X

Magneto-Acoustic Wave Motions in Visco-Elastic Thermal Memory Conductors

The magneto-acoustic problem with application to sound wave propagation through electrically conducting fluids, such as the earth's outer core or ocean in the earth's magnetic field or liquid metals has been addressed. This focuses on a continuum derivation of equations for total mass, linear momentum, and energy balances, as well as Maxwell's electrodynamic equations in the form of an induction equation. The theory can accommodate a thermal conductor with memory using Catanneo's assumption and thermodynamic equations of state that account for the magnetic field on the system.

The balance equations are field equations written for,

A. Electromagnetic Field Equations

Using Giorgi mks rationalized units,

$$\nabla \times H = j + \frac{\partial D}{\partial t} \qquad j = \sigma\left(E + v \times B - \varsigma(j \times B)\right)$$

$$\nabla \times E = -B_{,t}$$

$$\nabla \bullet B = 0$$

$$\nabla \bullet D = \rho_0 \qquad D = \varepsilon_0 E + P, \qquad H = \frac{B}{\mu_0}$$

j is the conduction current vector and ρ_0 represents the free charge density. Here it is assumed that the electrical properties do not change with density or compression (such as electro- or magnetostrictive materials). The magnetic field strength neglects the magnetization due to circular motion of charges, as well as that due to intrinsic magnetic moments. The polarization P, which is caused by both an apparent charge density from the gradient of polarization and a nonneutrality of charged particles (i.e., dipole moments), is assumed negligible for the acoustic materials of interest. Using the Maxwell equations for electrodynamics, the following induction equation with Ohm's law is formed. This presentation assumes a nonpolar material that neglects magnetization. The expression $1/(\mu\sigma)$ denotes a magnetic "viscosity" or diffusivity.

$$\mathbf{B}_t = \nabla\times(\nabla\times\mathbf{B})/(\mu\sigma) - \nabla\times(v\times\mathbf{B}) + \nabla\times[\varsigma/\mu(\nabla\times B)\times\mathbf{B}]$$

For μ and σ assumed constant, $\varsigma = 0$, and no ion slip,

$\mathbf{B}=\mathbf{B}_0+\mathbf{B}^a$, obtains $\mathbf{B}^a_t = (\nabla^2\,\mathbf{B}^a)/(\mu\sigma) + \nabla\times(v^a\times\mathbf{B}_0)$

B. Total Mass Balance

For no external sources of mass, the balance equation in the x-direction obtains,

$$\frac{\partial\rho}{\partial t} + v_x\frac{\partial\rho}{\partial x} = 0$$

C. Linear Momentum

The balance of total linear momentum is shown as Navier-Stokes' equations modified for a magneto-hydrodynamic fluid:

$$\rho\left(\frac{\partial v}{\partial t} + v\frac{\partial v}{\partial x}\right) = \frac{\partial s_{xx}}{\partial x} + (\underline{j}\times\underline{B})_x + \rho_e\underline{E}$$

with total stress tensor $\underline{S} = -p\,\underline{I} + \underline{T}$ and extra-stress $\underline{T}$.

In one dimension, $T_{xx} = E_0 \frac{\partial u}{\partial x} + E_1 \frac{\partial^\alpha}{\partial t^\alpha} \frac{\partial u}{\partial x}$

the stress tensor represents a relatively large class of nonlinear fluids with memory to first-order. For a rotating system, the Coriolos contribution $-2\rho\, \mathbf{\Omega} \times \mathbf{v}$ would appear on the right-hand side of the linear momentum balance for typical geophysical application. The **E** is the electric field and **B** represents the magnetic field vector. The ρ_e is the charge density and σ the conductivity.

O(1):

$$\rho_0 \left(\frac{\partial v_x^a}{\partial t} + v_x^a \frac{\partial v_x^a}{\partial x} \right) = -\frac{\partial p}{\partial x} + E_0 \frac{\partial^2 u}{\partial x^2} + E_1 \frac{\partial^\alpha}{\partial t^\alpha} \frac{\partial^2 u}{\partial x^2} - \left(\frac{B_z^0}{\mu} \frac{\partial B_z}{\partial x} + \frac{B_y^0}{\mu} \frac{\partial B_y}{\partial x} \right)$$

Differentiating the above equation with respect to time and subtracting

$$c_0^2 / \gamma \frac{\partial^2 v_x^a}{\partial x^2}$$

yields

$$\frac{\partial^2 v^a}{\partial t^2} - \frac{c_0^2}{\gamma} \frac{\partial^2 v^a}{\partial x^2} = -\frac{\partial^2 p^a}{\partial t \partial x} + E_0 \frac{\partial^3 u}{\partial t \partial x^2} + E_1 \frac{\partial^{\alpha+1}}{\partial t^{\alpha+1}} \frac{\partial^2 u}{\partial x^2} + \left(\frac{B_z^0}{\mu} \frac{\partial^2 B_z}{\partial t \partial x} + \frac{B_y^0}{\mu} \frac{\partial^2 B_y}{\partial t \partial x} \right) - \frac{c_0^2}{\gamma} \frac{\partial^2 v^a}{\partial x^2}$$

D. Energy

The O(1) equation for the internal energy, which represents the difference of the total and kinetic energies is written first for a Fourier-conducting material,

$$\rho_0 c_P \frac{\partial \theta^a}{\partial t} - \theta_0 \beta_\theta \frac{\partial p^a}{\partial t} - k_\theta \frac{\partial^2 \theta^a}{\partial x^2}$$

For a heat flux q_i that incorporates memory (Catanneo's assumption) or a Taylor expansion of the temperature gradient at the present time:

$$q_i = -k_\theta \left(\frac{\partial \theta}{\partial x_i} - \tau \frac{d}{dt}\left(\frac{\partial \theta}{\partial x_i} \right) \right), \qquad q_i + \tau \dot{q}_i = -k_\theta \frac{\partial \theta}{\partial x_i}$$

This heat transfer model for one dimension alters the next to last term above as follows:

$$-\frac{\partial}{\partial x}\left(\frac{k_\theta}{\tau_c} \right) \int e^{\Delta t / \tau_c} \frac{\partial \theta^a}{\partial x} dt$$. The linearized x component of

$$(\underline{j} \times \underline{B})_x = B_3^0 b_3 + B_2^0 b_2$$ and

$$p^a \approx c_0^2 \rho^a + \beta_1 \rho^{a2} + \frac{\partial p}{\partial \eta} \eta^a + \beta_2 \eta^{a2} + \sum_{i=1,3} \frac{\partial p}{\partial B_i} b_i + \sum \beta_j b_j^2 +$$

mixed second differential terms. Nonlinear coefficients are defined as

$$\beta_1 = \frac{1}{2}\frac{\partial^2 p}{\partial \rho^2}, \quad \beta_2 = \frac{1}{2}\frac{\partial^2 p}{\partial \eta^2}, \quad \beta_3 = \frac{1}{2}\frac{\partial^2 p}{\partial b_1^2}, \quad \beta_4 = \frac{1}{2}\frac{\partial^2 p}{\partial b_2^2}, \quad \beta_5 = \frac{1}{2}\frac{\partial^2 p}{\partial b_3^2}$$

neglecting second-order entropy, extra stress, and magnetic field pressure variations, as if they may be considered one order higher in the approximation.

$$\frac{\partial\rho^{a}}{\partial\tau} = \left\{ \frac{-c_{0}}{\rho_{0}}\frac{\partial\rho^{a}}{\partial x} + \left(\frac{-\beta_{1}}{c_{0}}\right)\rho^{a}\frac{\partial\rho^{a}}{\partial x} + \frac{\partial p}{\partial\eta}\left(\frac{k_{\theta}}{2c_{0}^{2}\rho_{0}\theta_{0}\tau_{c}}\right)\int_{-\infty}^{t} e^{\frac{\Delta t}{\tau_{c}}}\frac{\partial\theta^{a}}{\partial x}\,dt \right.$$

$$\left. + \frac{E_{0}}{2c_{0}^{2}}\frac{\partial^{2}u}{\partial x^{2}} + \frac{E_{1}}{2c_{0}^{2}}\frac{\partial^{\alpha}}{\partial t^{\alpha}}\frac{\partial^{2}u}{\partial x^{2}} + \frac{1}{2c_{0}^{2}}\sum_{i=4,5}\frac{\partial p}{\partial B_{i}}b_{i} \right\}$$

References

Cambel, A. B. *Plasma Physics and Magnetofluid Mechanics*. New York: McGraw-Hill, 1963.

Landau, L. D., and E. M. Lifshitz. *Electrodynamics of Continuous Media*. New York: Pergamon Press, 1960.

Grad, H. *The Magnetohydrodynamics of Conducting Fluids*. Edited by D. Bershader. California: Stanford University Press, 1959.

Alfven, H. *Cosmical Electrodynamics*. London: Oxford University Press, 1950.

Dungey, J. W. *Cosmic Electrodynamics*. New York: Cambridge, 1956.

Banos, A. *Electromagnetic Phenomena in Cosmical Physics*. Edited by B. Lehnert. New York: Cambridge, 1958.

Sherman, A., and G. Sutton. *Engineering Magnetohydrodynamics*. New York: McGraw-Hill, 1965.

Fowler, A. C. *Mathematical Models in the Applied Sciences (Cambridge Texts in Applied Mathematics)*(1997).

Mueller, I., and T. Ruggeri. *Rational Extended Thermodynamics*. New York: Springer, 1998.

Prigogine, I., and R. Defay. *Chemical Thermodynamics*. Translated by D. H. Everett. London: Longmans Green, 1954.

Korsunskii, S. V. *Soviet Physics Acoustic* 36, no. 1 (1990).

Koeller, R. C. *Journal of Applied Mechanics* 51 (1984): 299.

Bagley, R. L., *American Institute of Aeronatics and Astronautics Journal* 27, no. 10 (1989): 1412.

Appendix A

A nonhomogeneous initial value fractional differential equation of practical application can be written and solved as follows. Further equations, solutions, and methods are given for example in references 1-2.

$$\frac{d^{\alpha} y(t)}{dt^{\alpha}} - \lambda y(t) = h(t) \qquad (t > 0)$$

$$\frac{d^{\alpha - k} y(t;t=0)}{dt^{\alpha}} = b_k \qquad k = 1, n$$

Laplace transformed,

$$s^{\alpha} Y(s) - \lambda Y(s) = H(s) + \sum_{k=1}^{n} b_k s^{k-1}$$

$$\therefore Y(s) = \frac{H(s)}{s^{\alpha} - \lambda} + \sum_{k=1}^{n} \frac{s^{k-1}}{s^{\alpha} - \lambda}$$

The solution introduces Mittag-Leffler functions

$$E_{\alpha,\beta} = \sum_{k=0}^{\infty} \frac{z^k}{\Gamma(\alpha k+\beta)}; \qquad (\alpha>0, \quad \beta>0)$$

which are sometimes motivated as the equivalent of the usefulness of the exponential function in integer-order ordinary differential equations.

$$y(t) = \sum_{k=1}^{n} b_k t^{\alpha-k} E_{\alpha,\alpha-k+1}\left(\lambda t^{\alpha}\right) + \int_0^t (t-\tau)^{\alpha-1} E_{\alpha,\alpha}\left(\lambda(t-\tau)^{\alpha}\right) h(\tau)\, d\tau$$

Next, examine the simple equation,

$$\frac{d^{1/2} f(t)}{dt^{1/2}} + af(t) = 0 \qquad (t > 0)$$

$$\frac{d^{-1/2} f(t;t=0)}{dt^{-1/2}} = C$$

Applying the Laplace transform obtains $\frac{C}{s^{1/2} + \alpha}$ with

$$f(t) = Ct^{-1/2} E_{\frac{1}{2},\frac{1}{2}}\left(-a\sqrt{t}\right).$$

Other useful functions for writing solutions to fractional differential equations are given by Y. N. Rabotnov,

$$\Re_{\alpha}\left(\beta,t\right) = t^{\alpha} \sum_{k=0}^{\infty} \frac{\beta^{k} t^{k(\alpha+1)}}{\Gamma\left(k+1\right)\left(1+\alpha\right)} = t^{\alpha} E_{\alpha+1,\alpha+1}\left(\beta t^{\alpha+1}\right)$$

as well as other definitions on fractional calculus differentials or integrals have been examined by Podlubny (ref. 1).

References

Podlubny, I. "Fractional Differential Equations." *Mathematics in Science and Engineering* 198. New York: Academic Press, 1999.

Miller, K. S., and B. Ross. *An Introduction to Fractional Calculus and Fractional Differential Equations*. New York: John Wiley and Sons, 1993.

Appendix: Additional References and Notes

1. Westerlund has suggested alternative constitutive equations for D and B using fractional calculus integrals:

$$D = \varepsilon E^{\left(\alpha - 1\right)}, \qquad B = \mu H^{\left(\alpha - 1\right)}, \qquad (\alpha - 1 < 0)$$

where the superscript denotes the differentiation/integration order.

Westerlund, S. Casualty Report, N. 940426, University of Kalmar (1994).
M. Caputo has investigated fractional calculus differentials and introduced models similar to the fractional calculus rheology model for the electric field and flux.

2. M. Giona and H. E. Roman have presented generalizations for diffusive processes, which account for the fractal dimension of the media.

Giona, M., and H. E. Roman. *Chemical Engineering Journal* 49 (1992): 1.
Giona, M., S. Gerbelli, and H. E. Roman. *Physica A* 191 (1992): 449.
Roman, H. E. *Physical Review E* 51, no. 6 (1995): 5422.

3. A method referred to as semi-integral and differential electroanalysis has achieved experimental validation of measuring concentrations and diffusion coefficients of electro-active species in simple practical systems near electrode surfaces. Calculations of semi-differentials and integrals were made as follows for recorded current time transients' values I(t), I(Δ), I(2Δ), . . . I(j, . . . I(t) Δ) or charge q(t) at evenly spaced time-intervals.

$$\frac{d^{-1/2} I(t)}{d^{-1/2}} = \sqrt{\frac{\Delta}{\pi}} \sum_{j=1}^{t/\Delta} \left[I(j\Delta) + I(j\Delta - \Delta) \right] \left[\sqrt{\frac{t}{\Delta} - j + 1} - \sqrt{\frac{t}{\Delta} - j} \right]$$

$$\frac{d^{1/2} q(t)}{d^{1/2} t} = \frac{2}{\sqrt{\pi \Delta}} \left\{ \frac{q(0)}{2} \sqrt{\frac{\Delta}{t}} + \sum_{j=1}^{t/\Delta} \left[q(j\Delta) - q(j\Delta - \Delta) \right] \left[\sqrt{\frac{t}{\Delta} - j + 1} - \sqrt{\frac{t}{\Delta} - j} \right] \right\}$$

Grenness, M., and K. O. Oldham. *Analytical Chemistry* 44, no. 7 (June 1972).
K. B. Oldham, *Analytical Chemistry* 44, 196 (1972)
Goto, M., and D. Ishii. *Journal of Electroanalytical Chemistry and Interfacial Electrochemistry* 61 (1975): 361.
Goto, M., and K. B. Oldham. *Analytical Chemistry* 45, no. 12 (1973): 2043.
Goto, M., and K. B. Oldham. *Analytical Chemistry* 46, no. 11 (1973): 1522.
Goto, M., and K. B. Oldham. *Analytical Chemistry* 48, no. 12 (1976): 1671.

4. A well-known simple example of a fractional-order equation called Abel's integral equation. $\frac{1}{\Gamma(\alpha)}\int_0^{\tau}\frac{\varphi(\tau)d\tau}{(t-\tau)^{1-\alpha}} = f(t),\ \ t > 0$. The solution is written $\varphi(t)=\frac{1}{\Gamma(1-\alpha)}\frac{d}{dt}\int_0^{t}\frac{f(\tau)d\tau}{(t-\tau)^{\alpha}}$ or in a fractional Calculus denotation, $\frac{d^{-\alpha}\varphi(t)}{dt^{-\alpha}} = f(t)$ where $\frac{d^{\alpha}f(t)}{dt^{\alpha}} = \varphi(t)$.

   Abel, N. H. *Oeuvres Completes de Niels Henrik Abel* 1, 97.

5. Rabotnov, Yu. N. *Elements of Hereditary Solids*. Moscow: Nauka, 1977.

6. Extended thermodynamics has been applied to develop balance equations for relativistic magnetizable fluids.
   Palumbo, A. "A Model of Relativistic Magnetizable Fluid." Istituto del Biennio della Faclta di Ingegneria, Italy.

Nomenclature

B	Magnetic flux density [$MT^{-1}q^{-1}$]
D	Dielectric displacement [$L^{-2}q$]
E	Electric field intensity [$MLT^{-2}q^{-1}$]
f	frequency of wave [T^{-1}]
H	Magnetic field intensity [$L^{-1}T^{-1}q$]

J	Current density $[L^{-2}T^{-1}q]$
O( )	Order symbol
p, p^a	thermodynamic and acoustic pressures $[ML^{-1}T^{-2}]$
q	Coulomb Charge
$\underline{\mathbf{S}}$, $\mathbf{S}_{ij}$, $\underline{\mathbf{T}}$, T_{ij}	Total and extra-stress tensors and components $[ML^{-1}T^{-2}]$
E	inductive capacity $[M^{-1}L^{-3}T^2q^2]$
θ, θ^a	absolute and acoustic temperatures $[\theta]$
ρ, ρ^a	Total and acoustic density $[ML^{-3}]$
ρ_e	charge density $[L^{-3}q]$

[ ] means "dimensions of" {1} means "dimensionless" quantity; M = mass, L = Length, T = time, θ = temperature, H = heat (cal), mol= gmol.
(p)′ ≡ $\partial p/\partial t$ = differential of p with respect to t

XI

Acoustic Waves In Poro-elastic Materials

A. The poro-elastic equations below may be rewritten to a system of ordinary differential equations:

$$\begin{pmatrix} \rho_{11} & \rho_{12} \\ \rho_{21} & \rho_{22} \end{pmatrix}\begin{pmatrix} \phi_s \\ \phi_f \end{pmatrix} + \kappa\begin{pmatrix} 1 & -1 \\ -1 & 1 \end{pmatrix}\begin{pmatrix} \phi_s \\ \phi_f \end{pmatrix}_{t^{3/2}} - \begin{pmatrix} P & Q \\ Q & R \end{pmatrix}\begin{pmatrix} \phi_s \\ \phi_f \end{pmatrix}_{xx} = \begin{pmatrix} 0 \\ 0 \end{pmatrix} \qquad \underline{\phi} = \begin{pmatrix} \phi_s \\ \phi_f \end{pmatrix},$$

defining matrices $\underline{\underline{R}} = \begin{pmatrix} \rho_{11} & \rho_{12} \\ \rho_{21} & \rho_{22} \end{pmatrix}$ $\quad \underline{\underline{K}} = \kappa\begin{pmatrix} 1 & -1 \\ -1 & 1 \end{pmatrix}$ $\quad \Lambda = \begin{pmatrix} P & Q \\ Q & R \end{pmatrix}$

$$\underline{\underline{R}}\underline{\phi}_{tt} + \underline{\underline{K}}\underline{\phi}_{t^{3/2}} - \underline{\underline{\Lambda}}\underline{\phi}_{xx} = \underline{0}$$

transforming coordinates to a wave frame:

$$J \equiv x - ct, \qquad \frac{\partial}{\partial x} = \frac{\partial}{\partial J}, \qquad \frac{\partial}{\partial t} = -c\frac{\partial}{\partial J}.$$

In the transformed space, the equations of poro-elastic motion obtain,

$\left(c^2\underline{\underline{R}} - \underline{\underline{\Lambda}}\right)\underline{\phi}_{JJ} + c^{3/2}\underline{\underline{K}}\underline{\phi}_{J^{3/2}} = \underline{0}$. Let $\underline{\underline{M}} = \left(c^2\underline{\underline{R}} - \underline{\underline{\Lambda}}\right)$

$$\underline{\phi}_{JJ} + c^{3/2}\underline{\underline{M}}^{-1}\underline{\underline{K}}\underline{\phi}_{J^{3/2}} = 0$$

Let $\underline{\underline{Q}}\underline{\Psi} = \underline{\phi}$ $\qquad \underline{\Psi}_{JJ} + c^{3/2}\underline{\underline{Q}}^{-1}\underline{\underline{M}}^{-1}\underline{\underline{K}}\underline{\underline{Q}}\underline{\Psi}_{J^{3/2}} = 0$

For diagonal matrix $\underline{\underline{D}} = c^{3/2}\underline{\underline{Q}}^{-1}\underline{\underline{M}}^{-1}\underline{\underline{K}}\underline{\underline{Q}} = \begin{pmatrix} D_{11} & 0 \\ 0 & D_{22} \end{pmatrix}$

the equations may be separated,

$$\underline{\Psi_{1}}_{JJ} + D_{11}\Psi_{1_{J^{3/2}}} = 0$$

$$\underline{\Psi_{2}}_{JJ} + D_{22}\Psi_{2_{J^{3/2}}} = 0$$

Solving yields, $\Psi_i = m_i t^{-\frac{1}{2}} E_{\frac{1}{2},\frac{1}{2}}\left(-D_{ii}\sqrt{t}\right) \qquad i = 1,2$

Constants m_i are found from the boundary constants.

An alternative solution for plane sinusoidal waves of the form proportional to $\exp[Xx + i\omega t]$ for the complex propagation coefficient X, $X = -(\alpha + ik)$ written in terms of attenuation α and wave number k obtains a biquadratic equation in X.

$$cX^4 + dX^2 + e = 0$$

$$c = PR - RQ^2; \qquad d = a_{21}Q + a_{12}Q - a_{11}R - a_{22}P; \qquad e = a_{11}a_{22} - a_{21}a_{12}$$

$$\begin{pmatrix} a_{11} & a_{12} \\ a_{21} & a_{22} \end{pmatrix} = \begin{pmatrix} -\omega^2\rho_{11} + \kappa(i\omega)^{\frac{3}{2}} & -\omega^2\rho_{12} - \kappa(i\omega)^{\frac{3}{2}} \\ -\omega^2\rho_{21} - \kappa(i\omega)^{\frac{3}{2}} & -\omega^2\rho_{22} + \kappa(i\omega)^{\frac{3}{2}} \end{pmatrix}$$

$$P = \frac{(1-\phi)(1-\phi-(K_b/K_s))K_s + \phi(K_s/K_f)K_b}{1-\phi-(K_b/K_s)+\phi(K_s/K_f)} + \frac{4}{3}N$$

$$Q = \frac{(1-\phi-(K_b/K_s))\phi K_s}{1-\phi-(K_b/K_s)+\phi(K_s/K_f)}$$

$$R = \frac{\phi^2 K_s}{1-\phi-(K_b/K_s)+\phi(K_s/K_f)}$$

where $K_s = \dfrac{E_s}{3(1-2\nu_s)} \qquad N = \dfrac{E_b}{2(1+\nu_b)}$

E_s Young's modulus (solid)

E_b Young's modulus (porous skeletal frame)

ν_s Poisson coefficient (solid)

ν_b Poisson coefficient (skeletal frame)

$$\rho_{11} + \rho_{12} = (1-\phi)\rho_s$$
$$\rho_{12} + \rho_{22} = \phi\rho_f$$
$$\rho_{12} = -\phi\rho_f(\alpha - 1)$$

References

Sebaa, N., Z. E. A. Fallah, W. Lauriks, and C. Depollier. "Application of Fractional Calculus to Ultrasonic Wave Propagation in Human Cancellous Bone." Signal Processing, 2006.

Biot, M. A. *Acoustics, Elasticity, and Thermodynamics of Porous Media: Twenty-one Papers.* Edited by Ivan Tolstoy. New York: AIP Press, 1992.

B. Waves through a Semilinear (Nonlinear) Biot Poro-elastic

In vector matrix notation, the dynamical equations used by Donskoy may be written,

$$\underline{\underline{A}}\,\underline{V}_{xx} - \underline{\underline{B}}\,\underline{V}_{tt} = \left[\underline{\underline{C}}u_x + \underline{\underline{D}}w_x\right]\underline{V}_{xx}, \qquad \underline{V} = \begin{pmatrix} u \\ w \end{pmatrix}.$$

Transform to a wave frame that also helps to generate a system of ordinary differential equations using, $J = x - ct$, $\frac{\partial}{\partial x} = \frac{\partial}{\partial J}$, $\frac{\partial}{\partial t} = -c\frac{\partial}{\partial J}$,

$\underline{\underline{M}}\,\underline{V}_{JJ} = \underline{f}$ the matrix $\underline{\underline{M}} = \left(\underline{\underline{A}} - c^2\underline{\underline{B}}\right)$.

Solving the homogeneous system obtains u^L, v^L which are used to estimate the nonlinear right-hand terms $\underline{f}$. $\underline{f}^L \equiv (\underline{\underline{C}} u_J^L + \underline{\underline{D}} w_J^L) V_{JJ}^L$ $\quad \underline{V} = \underline{a}_1 v_1 + \underline{a}_2 v_2$

where $\underline{a}_1 = \begin{pmatrix} a_1 \\ a_3 \end{pmatrix}$ $\quad \underline{a}_2 = \begin{pmatrix} a2 \\ a_4 \end{pmatrix}$ and $v_1 = e^{i(\omega t - k_1 x)}$ $\quad v_1 = e^{i(\omega t - k_2 x)}$

Substituting these harmonic waveforms yields a determinant equation obtaining solutions for k_1, k_2 from solving $\left|\omega^2 \underline{\underline{B}} - k^2 \underline{\underline{A}}\right| = 0$.

$$\underline{V}^N = \underline{b} e^{i(\omega t - k_1 x)} e^{i(\omega t - k_2 x)} \quad \underline{b} = \begin{pmatrix} a_5 \\ a_6 \end{pmatrix} \quad \underline{\underline{M}}\,\underline{b}'' + 2ik\underline{\underline{M}}\,\underline{b}' - k^2 \underline{b} = \underline{f}^L e^{-kJ}, \quad k = k_1 + k_2$$

The system matrices are $\underline{\underline{A}} \equiv \begin{pmatrix} \lambda_c + \alpha^2 M & \alpha M \\ \alpha M & M \end{pmatrix}$ $\quad \underline{\underline{B}} \equiv \begin{pmatrix} \rho & \rho_f \\ \rho_f & M \end{pmatrix}$

$$\underline{\underline{C}} = \begin{pmatrix} 2(\alpha - 1)(\alpha^2 \beta^2 M^2 K - 2\alpha\phi MP + D) & c_{12} \\ 2M(3\alpha^2 \beta^2 M^2 K - 2\alpha\beta MK + P) & c_{22} \end{pmatrix}$$ with

$$c_{12} = 2\beta M(\alpha - 1)(\alpha\beta M)K - P \quad c_{22} = 2\beta M^2 K(3\alpha\beta M - 1)$$

$$\underline{\underline{D}} \equiv \begin{pmatrix} c_{12} & 2\beta^2 M^2 K(\alpha M - 1) \\ c_{22} & 6\beta^2 M^2 K \end{pmatrix} \qquad K = D + 6F + G, \qquad P = D + 2F$$

$$\alpha(\beta) \equiv 1 - \frac{K_s(\beta)}{K_r} \qquad M(\beta) = \frac{K_r^2}{K_r\left[1 + \beta\left(\frac{K_r}{K_f} - 1\right)\right] K_s(\beta)}$$

References

Donskoy, D. M., K. Khashanah, and T. G. McKee, T. G. "Nonlinear Acoustic Waves in Porous Media in the Context of Biot's Theory." *Journal of Acoustical Society of America* 102, no.5, Pt.1 (1997): 2521-2528.

XII

Sound Wave Propagation through Dilute Suspension of Particles

Abstract

The hydrodynamic or continuum approach is utilized to examine sound wave propagation through a dilute suspension of spherical particles in a viscous, heat-conducting fluid. The acoustical theory accounts for coupled phenomena among the phases, including mechanical particle-fluid interactions, such as Stokes's drag, virtual mass acceleration, and Basset (acceleration history) terms, as well as interfacial thermal energy transfer. Additional coupled phase phenomena, collectively called phoresis effects, are derived from modern continuum mechanics to account for particle transport due to gradients of temperature, density, or concentration (e.g., processes of thermophoresis, pcynophoresis, and diffusion phoresis). The role of these multiphase phoresis effects, as well as Stokes's drag are analyzed. Linearized volume-averaged balance equations for mass, linear momentum, and energy are solved for a plane wave of arbitrary frequency. Approximations are provided to enable better physical interpretation of the results and to compare to the earlier treatment by S. Temkin and R. A. Dobbins [*J. Acoust. Soc. Am.*, 40, 317-324, 1966] for an inviscid fluid phase but with a Stokes's drag force on each particle. The investigation also considers several generalizations for the case when the phoresis terms can be neglected. For example, a distribution of particle sizes is accounted for by developing a frequency-dependent function, which weights the drag forces by a particle-size distribution function. Furthermore, by invoking the correspondence principle, the drag force function for a Newtonian fluid is extended to a viscoelastic particle-laden material by using complex viscosities for shear and compressional relaxation functions. In the limit that the concentration of particles goes to zero, and the viscosity is Newtonian, the classical Kirchoff-Langevin equation is obtained. Several calculated results are provided for comparison to available experimental measurements, and a visco-elastic fluid suspension simulation illustrates attenuation and dispersion relationships versus particle size and concentration.

Introduction

The practical importance of acoustic wave propagation through multiphase media, such a dusty gases, fog, mist, sediment-laden water, aerosols, emulsions, and porous media has motivated a considerable amount of theoretical and experimental work. The theory developed here follows the hydrodynamic method used by Temkin and Dobbins and Marble for plane acoustic waves propagating through a dilute, homogeneous suspension of small (compared to the wavelength of the acoustic wave) spherical particles in a viscous fluid. We have extended earlier analyses to include explicitly the potential effects of thermophoresis, pycnophoresis, and barophoresis. Phoresis refers to the phenomena whereby particles are transported by gradients of temperature, concentration, pressure, or external forces such as electrophoresis. These coupled multiphase phenomena are derived from modern continuum mechanics for particle transport, and heretofore their role in acoustics has not been examined in addition to the familiar drag force calculated by Stokes.

This investigation also considers “non-Stokesian” drag forces in its treatment of mechanical particle-fluid interaction, as well as visco-thermal properties of the surrounding fluid. That is, both virtual mass acceleration and Basset time-history acceleration terms are included in the linear momentum exchanges between phases whenever phoresis terms can be neglected. Finally, generalization to a visco-elastic fluid case is made, and calculations were performed to apply the theory to estimate attenuation and sound speed versus frequency for several data sets available in the literature.

The results of the theory can be useful in order to (1) understand constitutive theories for coupled processes for colloid or particle motions as described by continuum mechanics; (2) establish relations for the sound speed and attenuation in terms of the measurable parameters of the dispersed system, such as the volume fraction of the particles, particle size, etc., which can then be used with online acoustic instrumentation in chemical processes or particulate contamination monitoring and controls or in material property measurement; and (3) design composite particulate systems having suitable acoustical properties.

Balance Equations

Formulation of the continuum macroscopic balance equations for multiphase flows has taken several modeling approaches such as averaging the field equations over appropriate time and/or space domains. The different averaging procedures have been discussed in several excellent reviews. Here we present the formulation of conservation and balance equations for multiphase flows, utilizing the volume-averaging procedure of Dobran.

In general, the interchanges of momentum and energy, and possibly mass, among the different phases are not easily specified. For the case of a concentrated suspension, the terms are quite complicated. However, for "dilute" suspensions, explicit relations can be obtained. Here we treat the case of dispersions that are sufficiently dilute such that the particles are noninteracting and only particle-continuous media interactions are considered. Then the effect on N-particles is the same as N-times the effect of one particle. This approach is the same as that used for the calculation of the effective shear viscosity (η_0) for a dilute dispersion of spherical Newtonian fluid particles in a Newtonian fluid, viz.,

$$\eta_0 = \bar{\eta}_C \left[1 + \phi_P \frac{\left(\bar{\eta}_C + \frac{5}{2} \bar{\eta}_P \right)}{\left(\bar{\eta}_C + \bar{\eta}_P \right)} \right]$$

where $\eta_C^\bullet$ and $\eta_P^\bullet$ are the shear viscosities of the continuous Newtonian phase and particles respectively. For $\eta_P^\bullet / \eta_C^\bullet$ «2 1,

$$\eta_0 = \bar{\eta}_C \left[1 + \frac{5}{2} \phi_P \right]$$

which is the Einstein formula for the apparent viscosity of a dispersion of solid particles and appears to be valid in the range of $\varphi < 0.02$ (2%). It is reasonable to expect that the dilute theory for acoustic propagation will have at least the same range of validity.

For a dilute suspension of rigid particles and neglecting surface tension, mass transfer, and chemical reaction at the interface, the balance equations for our planar wave propagation problem (in the x-direction of a rectangular Cartesian coordinate system) are written as,

$$\frac{\partial \bar{\rho}_C}{\partial t} + \frac{\partial (\bar{\rho}_C \tilde{V}_{Cx})}{\partial x} = 0 \tag{1}$$

$$\bar{\rho}_C \left[\frac{\partial \tilde{V}_{Cx}}{\partial t} + \tilde{V}_{Cx} \frac{\partial \tilde{V}_{Cx}}{\partial x} \right] = \frac{\partial \bar{S}_{Cxx}}{\partial x} + \bar{\rho}_C \; \tilde{b}_{Cx} + \overset{+}{p}_{Cx} \tag{2}$$

$$\bar{\rho}_\alpha \left[\frac{\partial \tilde{\varepsilon}_\alpha}{\partial t} + \tilde{V}_{\alpha x} \frac{\partial \tilde{\varepsilon}_\alpha}{\partial x} \right] = \mathbf{tr} \; (\bar{S}_\alpha \bar{D}_\alpha) + \mathrm{div} \; \bar{q}_\alpha + \bar{\rho}_\alpha r_\alpha + \overset{+}{\varepsilon}_\alpha \qquad \alpha = C, P \tag{3}$$

$$\bar{\rho}_P \left[\frac{\partial \tilde{V}_{Px}}{\partial t} + \tilde{V}_{Px} \frac{\partial \tilde{V}_{Px}}{\partial x} \right] = \frac{\partial \bar{S}_{Pxx}}{\partial x} + \bar{\rho}_P \; \tilde{b}_{Px} + p_{Px} \tag{4}$$

$$\frac{\partial \bar{\rho}_C}{\partial t} + \frac{\partial (\bar{\rho}_C \tilde{V}_{Cx})}{\partial x} = 0 \tag{5}$$

Equations (2) and (5) represent balances of mass for the continuous fluid mediums (variables denoted by subscript C) and particles (with subscript P), respectively. Equations (2) and (4) correspond to individual phase linear momentum equations with supply terms:

$$\overset{+}{p}_{C,x} \quad \text{and} \quad \overset{+}{p}_{Px}$$

while Eq. (3) is the balance equation for the total (e.g., for both phases) specific internal energy. These equations are constructed from volume averaging the macroscopic conservation laws for each phase and are consistent with the theory of multicomponent mixtures for a single phase in the limit that the interfacial area tends to zero. The partial density of phase α, $_\alpha$ is defined as $_\alpha = \varphi_\alpha \langle \rho_\alpha \rangle$ where the brackets around a field quantity indicate a phase average, and the overbar indicates a volume average:

$$\overline{F}_\alpha \equiv \frac{1}{V} \sum_{\delta} \int_{V_{\alpha\delta}} F_{\alpha\delta}\, dv \equiv \phi_\alpha < F_\alpha >$$

Here, $V_{\alpha\delta}$ is the volume of the δ-phase, and φ_α represents the volumetric fraction of the total volume occupied by each phase. The density of the dispersion $\rho = \overline{\rho}_C + \overline{\rho}_P$. Also, the density-weighted averaging operator for the velocity is defined by,

$$\tilde{V}_\alpha \equiv \phi_\alpha < \rho_\alpha V_\alpha > / \overline{\rho}_\alpha = \overline{\rho_\alpha \overline{V}_\alpha} \,/\, \overline{\rho}_\alpha$$

II. Constitutive Equations

A. Stress Tensor

Constitutive equations are required to complete the system of balance equations in order to derive the acoustic equations for a particular class of materials. In particular,

$$\overline{S}_{Cxx} = - p_C + \overline{T}_{Cxx} \; ; \; \overline{S}_{P\,xx} = - p_P + \overline{T}_{P\,xx}$$

The extra-stress tensors, (*xx*-component) for the continuous and the "particle" phases, which are treated as compressible linear materials are given by,

$$\overline{T}_{Cxx} = \beta_{CP} \frac{\partial \tilde{v}_{Px}}{\partial x} + \beta_{CC} \frac{\partial \tilde{v}_{Cx}}{\partial x}$$
$$\overline{T}_{Pxx} = \beta_{PP} \frac{\partial \tilde{v}_{Px}}{\partial x} + \beta_{PC} \frac{\partial \tilde{v}_{Cx}}{\partial x}$$

These stress-tensor terms that couple the phases would not appear in a constitutive theory that invokes *a priori,* the principle of phase separation. Acoustic viscosities are defined in terms of a linear combination of shear and bulk viscosities as in the ordinary Newtonian fluid case:

$$\beta_{\alpha\beta} \equiv K_{\alpha\beta} + \frac{4}{3} \eta_{\alpha\beta}$$

Experimental values for all these individual viscosity coefficients are lacking; however, explicit estimates were made for similar terms in a multicomponent theory developed by Goldman and Sirovich. We expect the results apply to solid particles if we allow the "viscosity" of the fluid particles to get large. This approach appears to be correct for predicting the viscosity of dispersion of solid particles in a continuous fluid medium (Einstein's formula).

The total stress tensor is defined by,

$$\bar{S} = \sum_{\alpha=1}^{\wp} \left(\bar{S}_{\alpha} - \bar{\rho}_{\alpha} \; \tilde{u}_{\alpha} \otimes \tilde{u}_{\alpha} \right)$$

C. Particle-fluid Interaction Terms (Newtonian fluid)

The interchange of linear momentum (and energy) between the particles and the continuous fluid matrix must be modeled in order to determine

$$\overset{+}{p}_{c} \quad \text{and} \quad \overset{+}{p}_{p} \; ,$$

respectively.

$$\sum_{\alpha=1}^{\wp} p_{\alpha}^{+} = 0 \quad \text{and} \quad \sum_{\alpha=1}^{\wp} \rho_{\alpha}^{+} = 0$$

Comparison to Marble, Tempkin, and Dobbins:

The balance equations developed by Marble and used by Temkin and Dobbins for a dispersion of rigid spherical particles in a perfect gas can be obtained from the two-phase equations (Eqs. 1-5) of Dobran if certain assumptions are made. It is noted that (1) the balance of mass for the particles matches if there is no interphase mass transfer; (2) the balance of mass for the continuous matrix matches if $\varphi_P \rightarrow 0$, hence $_P \rightarrow \rho^0_C$; (3) the balance of linear momentum for $\alpha = P$ matches if $S_p = 0$; and (4) the balance of linear momentum matches if stress tensors are chosen as,

$$S_{C\,xx} = -p + T^{a}_{P\,xx} \quad , \quad S_{P\,xx} = 0 \qquad (6)$$

with $\underline{\underline{T}}^{\bullet}_{Cxx}$ as the viscous extra-stress tensor for the pure Newtonian gas, i.e.,

$$\underline{\underline{T}}^{\Box}_{\underline{=}c} = 2\,\eta^{\Box}_{c}\overline{\underline{\underline{D}}}_{G} + (\,\overline{K}^{\Box}_{c} - 2\,\overline{\eta}^{\Box}_{c}\,/\,3\,)\,/\,(\,\text{tr}\,\overline{\underline{\underline{D}}}_{G})\,\underline{\underline{I}} \qquad (7)$$

where $\eta^{\bullet}_{C}$ and $K_{C}^{\bullet}$ are the shear and bulk viscosities of the pure gas respectively. It appears that the equations were formulated by requiring that the dispersion be dilute enough such that the particles are noninteracting, all particles in a local volume exhibit the same velocity vector and temperature.

III. Visco-elastic Fluids

Examples of visco-elastic media are biofluids, such as solutions of proteins, DNA, polysaccharides and glycoproteins, polymeric melts and solutions, and alcohols at low temperatures. In fact, most fluids are visco-elastic within an accessible range of acoustic frequencies. Therefore, it is desirable to extend the theory for acoustic waves propagating through particle-laden Newtonian fluids to the case of visco-elastic fluids.

A. Constitutive equation

The particular constitutive relation chosen here to represent the dependence of the stress tensor **S** on the motion of the material is that developed by Coleman and Noll for a *simple fluid,* which is a special case of the class of materials with memory:

$$\underline{\underline{S}}(\wp,t) = \overset{\infty}{\underset{\sigma=0}{S}}\,\big[\,C^{t}_{t}(\sigma);\rho(t)\,\big]$$

where $\wp$ represents the particle located at position x at time t; Θ is a functional whose argument is the entire history of the relative right Cauchy-Green tensor $C^{t}_{t}(\sigma)$ from present time t (here let τ be a time in the past $\tau \le t$, and let $\sigma \equiv t - \tau$, the past time) to the time in the very distant past ($\tau \to -\infty$ or $\sigma \to \infty$) and $\rho(t)$ is the density associated with $\wp$ at the time t. The relative right Cauchy-Green tensor is defined by,

$$\underline{\underline{C}}^{t}_{t}(\sigma) = \underline{\underline{F}}^{T}\,\underline{\underline{F}}$$

Let $\boldsymbol{X}$ denote the position of the particle $\wp$ in a reference state, then $\mathbf{F} \equiv \partial \boldsymbol{x}/\partial \boldsymbol{X}$ = grad $\boldsymbol{\chi}(\boldsymbol{X},\boldsymbol{\sigma})$. $\chi(X,\sigma)$ is called the relative deformation function and denotes the place or position of the particle $\wp$ at the time σ.

The functional [•] can be approximated for those motions such that the displacement of fluid particles is small by,

$$\underline{\underline{S}}(x,t) + p\,\underline{\underline{I}} \equiv \underline{\underline{T}}(x,t) = 2\int_{\tau=-\infty}^{t} \psi(t-\tau)\,\underline{\underline{D}}(x,\tau)\,d\tau + \underline{\underline{I}}\int_{\tau=-\infty}^{t} \phi(t-\tau)\,\mathrm{tr}\,\underline{\underline{D}}(x,\tau)\,d\tau$$

where $\underline{\underline{\mathbf{T}}}(\mathbf{x},\mathbf{t})$ is the extra-stress tensor, $\psi(\sigma)$ and $\varphi(\sigma)$ are the stress relaxation and the compressional relaxation functions respectively. Note that for $\psi(t\text{-}\tau) = \eta_0\,\delta(t\text{-}\tau)$ and $\varphi(t\text{-}\tau) = \lambda\,\delta(t\text{-}\tau)$, where $\delta(t\text{-}\tau)$ is the Dirac delta function centered at $\tau = t$. The above equation reduces to the Newtonian constitutive relation:

$$\underline{\underline{T}}(x,t) = 2\eta_0\,\underline{\underline{D}}(x,t) + \lambda_0\ I\ \mathrm{tr}\,\underline{\underline{D}}(x,t)$$

It is usually more convenient to define a bulk compressional relaxation function K(σ) as,

$$K(\sigma) \equiv \phi(\sigma) + \frac{2}{3}\psi(\sigma)$$

For a pure Newtonian fluid: $K(\sigma) = K_{CC}\,\delta(\sigma)$ where $K_{CC} = K_0 = \lambda + 2\eta/3$.

The stress relaxation function is well-known and has been studied experimentally and theoretically. The bulk compressional relaxation function has not been given the same attention. In fact, acoustic measurements are the only known means to determine the compressional relaxation function.

For oscillatory motions, it is convenient to work with the Fourier-Laplace transforms of the relaxation functions:

$$\eta^*(\omega) \equiv \int_0^\infty \psi(s)\,e^{-is\omega}\,ds;\quad K^* \equiv \int_0^\infty K(s)\,e^{lis\omega}\,ds$$

The complex functions $\eta^*(\omega)$ and $K^*(\omega)$ are called the frequency-dependent complex shear viscosity and bulk viscosity respectively. These quantities are alternative representations of the memory of the fluid for small displacements (motions). Experimental data for $\eta^*(\omega)$ are readily available for many materials, although the range of frequencies may be limited.

We remark that the mechanical constitutive equation for a *simple fluid* depends on the temperature history. Here we assume that the stress tensor $\mathbf{S}(\wp,t)$ is a functional of $\mathbf{C}_t^t(\sigma)$ and an ordinary function of the temperature at time t. Further, we assume that the thermodynamics of the system is that of a fluid without thermal memory. Specifically, the internal energy does not depend on the temperature history. We also note that this approach can be extended to include visco-elastic fluids subjected to motions with constant straining histories.

Visco-elastic Stokes's Drag

The results of the calculations for the oscillatory drag of a sphere in a Newtonian fluid for small Reynolds numbers can be extended to a visco-elastic fluid by using the "correspondence" principle. Basically this states that the complex-valued fields obtained for an oscillatory motion of a visco-elastic fluid are the same as those obtained for a Newtonian fluid by replacing the shear viscosity $\eta_C^\bullet$ by the complex viscosity η^*. Therefore, the drag exerted by an oscillating fluid on a mobile sphere is given by Eq. (10) where $\eta_C^\bullet$. Also, S_N^2 is replaced by,

$$(S^*)^2 \equiv \frac{i\omega\bar{\rho}_C R^2}{\eta^*}$$

in Eq. (13). The limiting case of a quiescent ($V_{C\omega} = 0$) visco-elastic fluid was analyzed by Thomas and Walters, Frater, and King and Waters using the Laplace transform method. We analyzed the fluid dynamical problem for the limiting case of a fixed sphere with a uniform oscillating fluid and obtained Eq. (14) with $V_{P\omega} = 0$. These checks on the limiting cases confirm the validity of using the Mazur and Bedeaux generalization of the Faxen theorem for viscoelastic fluids.

In this part of the paper we provide results for sound attenuation and dispersion in a multiphase system of particles where motions are affected by Stokes's drag, as well as potentially, by phoresis terms. These effects are explicitly included in the linear momentum supply as,

$$\overset{+}{p} = -a\,\Delta\theta - b\,\Delta\bar{\rho}_P - c\,\Delta\bar{\rho}_C - P_D\,\tilde{u}$$

The coefficient *a, b,* and *c* are related to the thermophoresis, diffusiophoresis, and barophoresis coefficients. Here, we also generalize the heat flux vector and internal energy function by,

$$\bar{q} = -k_\theta\,\Delta\,\theta - k_D\,\tilde{u}\;;\;\;\tilde{\varepsilon} = \tilde{\varepsilon}^\wedge(\theta, \bar{\rho}_P, \bar{\rho}_C)$$

where $\mathbf{u} = v_P - v_C$ is a relative (diffusive) velocity; k_θ is the thermal conductivity for the dispersion; and k_D is the Dufour coefficient. The effective thermal conductivity k_θ of a dispersion has been studied theoretically and experimentally. The problem of finding the thermal conductivity for a composite, consisting of a matrix containing a random dispersion of spheres whose thermal conductivity is different from that of the matrix, has been theoretically considered by Chang, Yendler, and Acrivos; and the results were compared to experiments.

Using these constitutive forms, the linearized equation of acoustics yields,

$$\frac{\partial \bar{\rho}_C^a}{\partial t} + \bar{\rho}_C^0 \frac{\partial \tilde{v}_{Cx}^a}{\partial x} = 0$$

$$\bar{\rho}^0 c_v \frac{\partial \theta^a}{\partial t} + \bar{\rho}^0 \left(\frac{\partial \bar{\varepsilon}}{\partial \bar{\rho}_C}\right) \frac{\partial \bar{\rho}_C^a}{\partial t} + \bar{\rho}^0 \left(\frac{\partial \tilde{\varepsilon}^\wedge}{\partial \bar{\rho}_P}\right) \frac{\partial \bar{\rho}_P^a}{\partial t} = k_\theta \frac{\partial^2 \theta^a}{\partial x^2} - k_D \frac{\partial \tilde{v}_{Cx}^a}{\partial x} + k_D \frac{\partial v_{Px}^a}{\partial x}$$

$$\frac{\partial \bar{\rho}_C^a}{\partial t} + \bar{\rho}_C^0 \frac{\partial \tilde{v}_{Cx}^a}{\partial x} = 0$$

$$\bar{\rho}_C^0 \frac{\partial \tilde{v}_{Cx}^a}{\partial t} = -\frac{c_0^2}{\gamma} \beta_\theta \frac{\partial \theta^a}{\partial x} - \frac{c_0^2}{\gamma}(1 + \beta_C \bar{c}_C^0) \frac{\partial \bar{\rho}_P^a}{\partial x} - \frac{c_0^2}{\gamma}(1 - \beta_C \bar{c}_P^0) \frac{\partial \bar{\rho}_C^a}{\partial x}$$

$$+ P_D \tilde{u}_x^a + \beta_{CC} \frac{\partial^2 \tilde{v}_{Cx}^a}{\partial x^2} + \beta_{CP} \frac{\partial^2 \tilde{v}_{Px^2}^a}{\partial x} - a \frac{\partial \theta^a}{\partial x} - b \frac{\partial \bar{\rho}_P^a}{\partial x} - c \frac{\partial \bar{\rho}_C^a}{\partial x}$$

$$\bar{\rho}_P^0 \frac{\partial \tilde{v}_{Px}^a}{\partial t} = \beta_{PP} \frac{\partial^2 \tilde{v}_{Px}^a}{\partial x^2} + \beta_{PC} \frac{\partial^2 \tilde{v}_{Cx}^a}{\partial x^2} + P_D \tilde{u}^a - a \frac{\partial \theta^a}{\partial x} - b \frac{\partial \bar{\rho}_P^a}{\partial x} - c \frac{\partial \bar{\rho}_C^a}{\partial x}$$

Bicubic Characteristic Equation (Phoresis Effects)

Assuming damped harmonic wave solutions of the form:

$$(\bar{\rho}_C^a, \tilde{v}_C^a, \theta^a, \bar{\rho}_P^a, \bar{\rho}_C^a) = (\bar{\rho}_C^0, \tilde{v}_C^0, \theta^0, \bar{\rho}_P^0, \bar{\rho}_C^0) \exp(\chi x + i\omega t)$$

and substituting these into the linearized balance equations (Eqs. (xxxx)) obtains a set of algebraic equations for the magnitudes of the acoustic variables. Nontrivial solution for these quantities are obtained, provided that the determinant of their coefficients is zero, (det **M** = 0),

$$M_{11} = M_{14} = M_{15} = M_{41} = M_{42} = M_{43} = 0$$

$$M_{12} = M_{44} = i\omega \qquad M_{13} = \bar{\rho}_C^0 \chi$$

$$M_{21} = \frac{\bar{\rho}^0 c_0^2 \beta_\theta \chi}{\gamma}$$

$$M_{22} = \frac{c_0^2 (1 - \beta_C \bar{c}_P) \chi}{\gamma} - c\chi \quad M_{23} = i\omega \bar{\rho}_C^0 - \beta_{CC} \chi^2 + P_D \quad M_{24} = c_0^2 (1 + \beta_C \bar{c}_C^0) \chi - b\chi$$

$$M_{25} = -\beta_{CP} \chi^2 - P_D$$

$$M_{31} = -k_\theta \chi^2 + i\omega \bar{\rho}^0 c_v$$

where the components of **M** are given by,

$$M_{45} = \bar{\rho}_P \chi \quad ; \quad M_{54} = b\chi \quad M_{51} = a\chi \quad ; \quad M_{55} = i\omega \bar{\rho}_P^0 - \beta_{PP} \chi^2 + P_D$$

$$M_{52} = c\chi$$

$$M_{35} = i\omega\left(\rho^{0}\frac{\partial\breve{\varepsilon}}{\partial\bar{\rho}_{P}}\right);\quad M_{53} = -(P_{D} + \beta_{PC}\chi^{2})$$

$$M_{33} = -M_{35} = k_{D}\,\chi$$

$$M_{32} = i\omega\left(\bar{\rho}^{0}\frac{\partial\breve{\varepsilon}}{\partial\bar{\rho}_{C}}\right)$$

with

$$\bar{\rho}^{0}\frac{\partial\breve{\varepsilon}}{\partial\bar{\rho}_{P}} = \left[-\frac{c_{p} - c_{v}}{\beta_{\theta}} + \bar{c}_{C}^{0}\frac{\partial\hat{\varepsilon}}{\partial\bar{c}_{P}}\right]$$

$$\bar{\rho}^{0}\frac{\partial\breve{\varepsilon}}{\partial\bar{\rho}_{C}} = \left[-\frac{c_{p} - c_{v}}{\beta_{\theta}} - \bar{c}_{P}^{0}\frac{\partial\hat{\varepsilon}}{\partial\bar{c}_{P}}\right]$$

and

$$\frac{\partial\hat{\varepsilon}}{\partial\bar{c}_{P}} = \frac{\partial\hat{h}}{\partial\bar{c}_{P}} - \frac{\theta\,\beta_{\theta}\,\beta_{C}\,c_{0}^{2}}{\gamma}$$

Here, the specific enthalpy is $\bar{h} = \tilde{h} = \breve{h}(\theta, p, c_{P})$; and the specific internal energy is $\bar{\varepsilon} = \breve{\varepsilon}\left(\theta, \bar{\rho}_{P}, \bar{\rho}_{C}\right) = \hat{\varepsilon}\left(\theta, p, \bar{c}_{P}\right)$; and we define

as the heat of mixing of particles added to the continuous phase.

$$H \equiv \frac{\partial\hat{h}}{\partial\bar{c}_{P}}$$

Dimensionless parameters are defined,

$\gamma = \dfrac{c_{p}}{c_{v}}$ Heat capacity ratio

$Y \equiv \dfrac{k_{\theta}}{\beta_{0}\,c_{p}}$ Thermoviscous number

$Z \equiv X\,Y = \dfrac{k_{\theta}\,\omega}{\rho c_{0}^{2}\,c_{p}}$ Thermal conduction frequency number

$$X_{\alpha\beta} \equiv \frac{\beta_{\alpha\beta}\,\omega}{\rho\, c_0^2} \qquad \text{Acoustic viscous frequency number}$$

$$X \equiv \sum_{\alpha,\beta} X_{\alpha\beta} = \frac{\beta_0\,\omega}{\rho\, c_0^2} \qquad \text{Overall acoustic viscosity frequency number}$$

$$A \equiv \frac{a}{\rho\,\beta_\theta\, c_0^2} \qquad \text{Thermophoresis number}$$

$$B \equiv \frac{b}{c_0^2} \qquad \text{Density\_phoresis number (particles)}$$

$$C \equiv \frac{c}{c_0^2} \qquad \text{Density\_phoresis number (continuous phase)}$$

$$P = \frac{P_D}{\rho\,\omega} \qquad \text{Frictional coefficient}$$

$$K_D \equiv \frac{k_D\,\beta_\theta}{\rho\, c_v\,\gamma} \qquad \text{Dufour number}$$

$$H \equiv \frac{H\,\beta_\theta}{c_v\,\gamma} \qquad \text{Heat\_of\_mixing number}$$

We have omitted all the superscripts, and it is understood that the physical quantities and variables are determined at the static values (ρ^0, θ^0, c^0).

Expansion of det **M** yields the bicubic characteristic equation:

$$\{-i(\mathbf{X}_{CC}\mathbf{X}_{PP} - \mathbf{X}_{CP}\mathbf{X}_{PC})\,\gamma\mathbf{Z}/(\bar{c}_C\bar{c}_P) + [-\mathbf{B}(\mathbf{X}_{PC} + \mathbf{X}_{CC}) + \mathbf{X}_{PC}(\mathbf{1}+\beta_C\,\bar{c}_C)\gamma^{-1}]Z\gamma/$$

$$\bar{c}_C + [\mathbf{C}(\mathbf{X}_{PP} + \mathbf{X}_{CP}) - \mathbf{X}_{PP}(1-\beta_C\,\bar{c}_P)\gamma^{-1}]\,Z\,\gamma/\,\bar{c}_P\ \}(i\,\bar{c}_P\,/P_D)\,(\chi/k_0)^6 + \{(-i[$$

$$(\gamma\mathbf{Z}\,\mathbf{X}_{CC}\,\bar{c}_P)/\,\bar{c}_C\ -(\mathbf{X}_{CP}\mathbf{X}_{PC} - \mathbf{X}_{CC}\mathbf{X}_{PP}\,/\,\bar{c}_C + \gamma\,\mathbf{Z}\,\mathbf{X}_{PP}\,]\,\{\,\beta_C\,\bar{c}_P\,[\,(\gamma-1)\,/$$

$$\gamma\,][\,\mathbf{X}_{PC}\mathbf{-A}(\mathbf{X}_{PC}+\mathbf{X}_{CC}) - \mathbf{A}(\mathbf{X}_{PP}+\mathbf{X}_{CP}) + \{\,-\mathbf{B}\,(\mathbf{X}_{PC}+\mathbf{X}_{CC})\,-\mathbf{X}_{PC}\,\bar{c}_C\ \mathbf{H} + \beta_C\,\bar{c}_C$$

$$\mathbf{X}_{PC}\,\gamma^{-1} + \mathbf{X}_{PC} + (\mathbf{C-B}) - (\mathbf{1}-\beta_C\,\bar{c}_P\,)\mathbf{Z}\,\bar{c}_C\ + \mathbf{A}(\mathbf{X}_{PC}+\mathbf{X}_{CC})\,\}\,(1-\bar{c}_P\,+\gamma\mathbf{H}-\gamma)\}\ \bar{c}_P\,/$$

$$\bar{c}_C\ \gamma + \{\,[\,(\mathbf{X}_{PP}+\mathbf{X}_{CP})\,\mathbf{C} - (1-\beta_C\,\bar{c}_P\ \gamma^{-1} + \gamma\,\bar{c}_P\ \mathrm{H})\,\mathbf{X}_{PP}]\ \bar{c}_C\ -(\mathbf{X}_{PP}+\mathbf{X}_{PC})\mathbf{K}_D$$

$$\gamma^{-1}\}/\,\bar{c}_C\ +[-\mathbf{A}(\mathbf{X}_{PP}+\mathbf{X}_{CP})\,(\mathbf{1-H}\,\bar{c}_P\ \gamma)]-\gamma\mathbf{XZ}\,P_D])\,/\,\bar{c}_C\ + \mathbf{AXK}_D\,/\,\bar{c}_C\ )\,/\,\mathbf{P}_D$$

$$+\,\beta_C\,(1-\gamma^{-1})\,(i\,\bar{c}_P\ /\,P_D\,)(\,\mathbf{A}\text{-}\mathbf{B}\text{-}\bar{c}_C\ \ \mathbf{C}) + \{\,[\,\mathbf{iH}\,(\bar{c}_C\ \ \mathbf{C} + \bar{c}_P\ \mathbf{B}\,) - (\,i\,\beta_C/$$

$\gamma)(\bar{c}_C\ \mathbf{C}+\bar{c}_P\ \mathbf{B})+i\,(\mathbf{B}-\mathbf{C})\,\bar{c}_C\ +i\,\mathbf{B}\,\mathbf{K}_D\,\gamma^{-1}-I\,\mathbf{A}\,[\,\mathbf{H}+(1-\gamma^{-1})\,\beta_C\,]\bar{c}_C$
$+i\,(\,1+\beta_C\,\bar{c}_C\ )\,\mathbf{Z}\,\mathbf{P}_D-i\,\mathbf{A}\,(\,1+\beta_C\,\bar{c}_C\ )\,\mathbf{K}_D\,\}\,(\bar{c}_P\,/\,\bar{c}_C\ \ \mathbf{P}_D)+i\,\mathbf{C}\,\mathbf{K}_D/\,\gamma$
$\mathbf{P}_D+i\,\gamma\,(\,1-\beta_C\,\bar{c}_P\ \ )\,\mathbf{Z}\text{-}(\,i\mathbf{A}\,\mathbf{K}_D/\,\gamma\,\mathbf{P}_D)\,(\,1-\beta_C\,\bar{c}_P\ \ )\}\,(\,\chi\,/\,k_0)^4+\{\,(\,1+\bar{c}_P$
$/\,\bar{c}_C\ \ )+i\,\bar{c}_P\,/\,\mathbf{P}_D)\,(\,1-\beta_C\,\bar{c}_P\ \ )+(\,i\bar{c}_P\ \ /\,\mathbf{P}_D\,)[\,\beta_C(\,\gamma-1\,)-\mathbf{H}\,\gamma\text{-}\mathbf{K}_D\,/\,(\bar{c}_P$
$\bar{c}_C\ )]\,\mathbf{A}+i\,(\mathbf{B}-\mathbf{C}\,)\,/\,\mathbf{P}_D+(i\,\bar{c}_P\,/\,\mathbf{P}_D)\,\bar{c}_P\ \mathbf{H}+i\,\mathbf{K}_D/\,\gamma\,\mathbf{P}_D$
$+(\,i\,\mathbf{X}/\,\bar{c}_C\ )[\,\mathbf{X}_{CC}/\mathbf{X}]\,i\,\bar{c}_P\,/\,\mathbf{P}_D+(\,i\,\bar{c}_C\,/\,\mathbf{P}_D)\,\mathbf{X}_{PP}\,/\,\mathbf{X}+1\,]+i\,\gamma\,\mathbf{Z}(\,1+\bar{c}_P\,/$
$\bar{c}_C\ +i\,\bar{c}_P\,/\,\mathbf{P}_D\,)\}$
$*(\,\chi\,/\,k_0)^2+[\,1+\bar{c}_P\,/\,\bar{c}_C\ \ +i\,\bar{c}_P\,/\,\mathbf{P}_D\,]=0$

For the case of a pure Newtonian fluid, this reduces to,

$$\left(\frac{\chi}{k_0}\right)^4[X\,Y]+\left(\frac{\chi}{k_0}\right)^2[1+i\,X\,(1+\gamma\,Y)\,]=1=0$$

which is the classical Kirchoff-Langevin biquadratic equation for a Newtonian fluid.

If, (1) the phoresis terms and the Dufour term are neglected ($\mathbf{A}=\mathbf{B}=\mathbf{C}=\mathbf{K_D}=0$); (2) the heat-of-mixing is negligible, then determinant equation becomes,

$$\{(1+\bar{c}_P\,/\,\bar{c}_C\,)+(i\bar{c}_P\,/\,P_D\,)(1-\beta_C\,\bar{c}_P\,)+(iX\,/\,\bar{c}_C\,)[1+X_{CC}\,(i\bar{c}_P\,/\,P_D\,)/\,X\,]$$
$$+\gamma\,XY\,(1+\bar{c}_P\,/\,\bar{c}_C+i\bar{c}_P\,/\,P_D\,)\}(\chi\,/\,k_0\,)^2\ +[1+\bar{c}_P\,/\,\bar{c}_C+i\bar{c}_P\,/\,P_D\,]=0$$

Dividing through by $(1+ic_P/P_D)$ obtains,

$$\{1+iX+iXY\gamma\left[1+\bar{c}_P\,/\left(1+i\bar{c}_P P_D\right)\right]+\left(\bar{c}_P\,/\,\bar{c}_C\right)\left[1-\beta_C\,\bar{c}_C\,(i\bar{c}_P\,/\,P_D)\right]/\left(1+i\bar{c}_P\,/\,P_D\right)\}+1+\bar{c}_P\,/\left(1+i\bar{c}_P\,/\,P_D\right)=0$$

For the case of a dilute dispersion $\bar{c}_P \prec\prec 1, \bar{c}_c \approx 1$ and the density ratio $R \equiv \rho_P^{\bullet}/\rho_C^{\bullet} >> 1$ (therefore $\beta_C \to 1$), the above equation obtains,

$$[1+\bar{c}_P\,)+iX\,(1+\gamma Y\,\Lambda_V\,\bar{c}_P\,)](\chi\,/\,k_0)^2 1+\bar{c}_P\Lambda_V=0,\quad \Lambda_V\equiv\left(1+i\bar{c}_P\,/\,P_D\right)^{-1}$$

If the momentum production is due primarily to Stokes's drag, then $P^D = n_p\zeta$, where n_p is the number density of particles; ζ is the "frictional coefficient", and $\zeta = 6\pi\eta_C{}^\bullet R$ for a spherical particle. Then

$$\Lambda_V = \Lambda_V^S \equiv \left(1 + \frac{2}{9} - S_N^2 \right)^{-1}$$

where $S_N{}^2$ is the Newtonian Stokes number defined by,

$$S_N^2 \equiv \frac{i\omega R^2 \bar{\rho}_C}{\bar{\eta}_C}$$

For an inviscid, continuous fluid (**X** = 0):

$$(\chi / k_O)^2 + \bar{c}_P \Lambda_V^S + 1 = 0$$

which is precisely the form obtained by Temkin and Dobbins, however lacking the terms representing the interchange of energy between the phases.

Details and extensions of this theory to accommodate other drag force terms such as virtual mass acceleration and Basset time-history for a distribution of particle sizes, as well as calculated examples with data derived from the literature can be found in reference 19.

References

Temkin, S., and R. A. Dobbins. *Journal of Acoustical Society of America* 40 (1966): 1016-1024.

—. *Journal of Acoustical Society of America* 40 (1966): 317-324.

Marble, F. *Combustion and Propulsion: Fifth AGARD Colloquium.* Edited by R. Hagerty, A. Jaunotte, O. Lutz, and S. Penner. New York: MacMillan, 1963.

Marble, F. *Annual Review of Fluid Mechanics* 2. Edited by M. Van Dyke, W. G. Vincent, and J. V. Wehausen. Annual Reviews 1970, 379-447.

F. Dobran. *International Journal of Multiphase Flow* 10 (1984): 273-305

—. *International Journal of Multiphase Flow* 11 (1985): 1-30.

Dobran, F. in "Multiphase Flow and Heat Transfer III, Part A: Fundamentals." Edited by T. Veziroglu and A. Bergles. *Elsevier Science* 1983, 23-39.

Drew, D. *Archive for Rational Mechechanics and Analysis* 62 (1976): 149-163.

Drew, D., and R. Lahey. *International Journal of Multiphase Flow* 5 (1979): 243-64.

Goldman, E., and L. Sirovich. *Physics of Fluids* 10, no. 9 (1967).

Bowen, R. M., and D. J. Garcia. *International Journal of Engineering Science* 8 (1970): 63-83.

Dunwoody, N., and I. Mueller. *Archive for Rational Mechanics and Analysis*. 344-69.

Margulies, T., and W. H. Schwarz. *Frontiers in Fluid Mechanics*. Edited by J. L. Lumley and S. H. Davis. 220-280. Berlin: Springer-Verlag.

Margulies, T., and W. H. Schwarz. *Journal of Acoustical Society of America* 82, no.2 (1987): 522-533.

Stokes, G. *Trans. Cambridge Philosophical Society* 9, no.2 (1851): 8.

Basset, A. B. *Philosophical Transaction of Royal Society of London* 179 (1888): 43-69.

—. *Quarterly Journal of Math* 41 (1910): 369-381.

Faxen, H. Arkiv för Matematik Astronomi och Fysik 18, no. 29 (1914): 1-52.

Batchelor, G. *Journal of Fluid Mechanics* 41 (1970): 545-70.

Saffman, P. *Journal of Fluid Mechanics* 22 (1965): 383-400.

Maxey, M., and J. Riley. *Physics of Fluids* 26 (1983): 883-889.

Margulies, T., and W. H. Schwarz. *Journal of Acoustical Society of America* 96, no.1 (1994): 319-331.

Schwarz, W. H., and T. Margulies. *Journal of Acoustical Society of America* 90, no 6 (1991): 3209-3217.

Davidson, G. A. *Aerosol Science* 5 (1974): 55-69.

XIII

Hydrothermal Vorticity Equations: With Equations of State

List of Symbols

ς	chemical affinity [ML^2T^{-2}]
c_0	speed of sound [LT^{-1}]
c_p	heat capacity at constant pressure [$L^2T^{-2}\theta$]
c_v	heat capacity at constant volume [$L^2T^{-2}\theta$]
g	gravitational acceleration [LT^{-2}]
Y	heat of reaction (at constant temperature and pressure) [H mol^{-1}]
p	thermodynamic pressures [$ML^{-1}T^{-2}$]
$\underline{\mathbf{S}}, \underline{\mathbf{T}}$	Total and extra-stress tensors and components [$ML^{-1}T^{-2}$]
t	time variable [T]
v	velocity vector [LT^{-1}]
Y	enthalpy of reaction
z	vertical position
θ	absolute temperatures [θ]
β_θ	isothermal coefficient of thermal expansion [θ^{-1}]
ρ	total density [ML^{-3}]
ξ	reaction progress variable [mol M^{-1}]
υ	specific volume = $1/\rho$ L^3M^{-3}]
$\underline{\omega}$	vorticity vector
[ ]	means "dimensions of" {1} means "dimensionless" quantity; M = mass, L = Length, T = time, θ = temperature, H = heat (cal), mol= gmol.

Introduction:

The local form of the field equations for the balances of total mass, linear momentum, and energy are formulated in ways to develop alternate useful forms of the Vorticity-Bernouilli equations. The engineering and physics applications of these equations are continually emphasized with applications and widespread use. [1-5] Here, extended developments of the equations are explored to provide more insight into their construction and generalization. Earlier work to generalize the these equations has been elaborated upon in ways that appear fruitful for analysis and for experimentation to illuminate a better understanding of fluid flow.

Derivation Using Local Field Equations:

The balance equation for the total mass without sources is written,

$$\frac{\partial \rho}{\partial \mathbf{t}} + \nabla \cdot (\rho \underline{\mathbf{v}}) = 0 \tag{1}$$

The differential operator for the total derivative $\frac{\mathrm{D}}{\mathrm{Dt}}$ is introduced for capturing the temporal and convective acceleration contributions,

$$\frac{\partial \rho}{\partial \mathbf{t}} + \rho \nabla \cdot \underline{\mathbf{v}} + \underline{\mathbf{v}} \cdot \nabla \rho = 0$$

$$\frac{\mathbf{D}\rho}{\mathbf{Dt}} + \rho \nabla \cdot \underline{\mathbf{v}} = 0 \quad \text{or} \quad \dot{\rho} + \rho \nabla \cdot \underline{\mathbf{v}} = 0 \quad , \quad -\nabla \cdot \underline{\mathbf{v}} = \frac{\dot{\rho}}{\rho}$$

Next the linear momentum equations, where $\underline{\underline{\mathbf{T}}}$ denotes the extra-stress tensor, and $\mathbf{F}_{\mathbf{B}}$ denotes the body force follow. They are presented using both the compact total derivative and the expanded form with its definition,

$$\rho \frac{\mathbf{D}\underline{\mathbf{v}}}{\mathbf{Dt}} = -\nabla \mathbf{p} + \nabla \cdot \underline{\underline{\mathbf{T}}} + \rho \underline{\mathbf{F}}_{\mathrm{b}}$$

$$\rho \frac{\partial \underline{\mathbf{v}}}{\partial \mathbf{t}} + \rho \underline{\mathbf{v}} \cdot \nabla \underline{\mathbf{v}} = -\nabla \mathbf{p} + \nabla \cdot \underline{\underline{\mathbf{T}}} + \rho \underline{\mathbf{F}}_{\mathrm{b}} \tag{2}$$

In a conservative body force situation such as a gravitational field one may define a potential function (per unit mass) ϕ, such that $\underline{F}_b = -\nabla\phi$.

Furthermore, a modified potential Π may be defined which appears useful in the constant density case.

$$\frac{D\underline{v}}{Dt} = -\frac{\nabla p}{\rho} + \frac{\nabla \cdot \underline{\underline{T}}}{\rho} - \nabla\varphi \quad \text{Then for constant density,}$$

$$\frac{D\underline{v}}{Dt} = -\nabla\Pi + \frac{\nabla \cdot \underline{\underline{T}}}{\rho} \qquad \textbf{where} \qquad \Pi \equiv \frac{p}{\rho} + \varphi$$

Recalling the vector identity, $\underline{v} \cdot \nabla \underline{v} = \nabla\left(\frac{1}{2}v^2\right) - \nabla \times \nabla \times \underline{v}$, the vorticity vector $\underline{\omega}$; in addition to integrating the compressible case along a streamline coordinate, here denoted by s yields,

$$\int_{s_1}^{s_2} \frac{\partial \underline{v}}{\partial t} \cdot ds + \int_{s_1}^{s_2} \nabla \frac{v^2}{2} \cdot ds - \int_{s_1}^{s_2} \nabla \times \underline{\omega} \cdot ds =$$
$$-\int_{s_1}^{s_2} \frac{1}{\rho} \nabla p \cdot ds + \int_{s_1}^{s_2} \frac{1}{\rho} \frac{1}{A} \int \nabla \cdot \underline{\underline{T}} \cdot \underline{n} dA \cdot ds - \int_{s_1}^{s_2} \nabla\varphi \cdot ds \tag{3}$$

If this situation corresponds to a no flow condition perpendicular to the streamline, this would imply that the vorticity is zero in the interior fluid. It is recognized that even in laminar flow parallel to a conduit wall vorticity is generated at the boundary arising from viscous stresses.

Now one may proceed to examine an equation of state for the pressure as a function of density ρ, and of temperature θ. This approach explicitly incorporates for the first differential of pressure an equation of state delivered from independent, separate effects experiments. Note

$$\int_{s_1}^{s_2} \frac{1}{\rho} \nabla p \cdot ds \quad \text{where} \quad dp = \nabla p \cdot ds \quad \text{obtains} \int_{s_1}^{s_2} \frac{dp}{\rho}$$

Consider as an example a non-reacting system and a Van der Waal's Equation of State. Let $dp(\rho,\theta) = \left.\frac{\partial p}{\partial \rho}\right|_{\theta} d\rho + \left.\frac{\partial p}{\partial \theta}\right|_{\rho} d\theta$ so that

$$\mathbf{p} = \frac{\mathbf{a}\rho^2}{\mathbf{M}_w^2} + \frac{\mathbf{R}\theta}{\left(\mathbf{M}_w/\rho - \mathbf{b}\right)} = \frac{\mathbf{a}\rho^2}{\mathbf{M}_w^2} + \frac{\rho\mathbf{R}\theta}{\mathbf{M}_w - \rho\mathbf{b}} \tag{4}$$

which obtains the following partial differentials,

$$\left.\frac{\partial \mathbf{p}}{\partial \rho}\right|_{\theta} = \frac{2\mathbf{a}\rho}{\mathbf{M}_w^2} - \frac{\rho\mathbf{b}\mathbf{R}\theta}{(\mathbf{M}_w - \rho\mathbf{b})^2}, \qquad \left.\frac{\partial \mathbf{p}}{\partial \theta}\right|_{\rho} = \frac{\rho\mathbf{R}}{(\mathbf{M}_w - \rho\mathbf{b})}$$

Formulating the integration of the differential pressure divided by density for several Equations of State including Tait's, Van der Waal's, Redlich-Kwong's equation, and Beattie-Bridgman's yields Table 1 for a hydrothermal systems viewpoint.

Table 1:

Equation of State	$\int \frac{\mathbf{dp}}{\rho}$
Tait's $\mathbf{p} = \mathbf{A}\left[\left(\frac{\rho}{\rho_0}\right)^{\mathbf{z}} - 1\right]$	$\frac{\mathbf{A}\rho^{\mathbf{z}}}{(\mathbf{z}-1)\rho_0^{\mathbf{z}}}$
Van der Waal's $\mathbf{p} = \frac{\mathbf{a}\rho^2}{\mathbf{M}_w^2} + \frac{\mathbf{R}\theta}{\left(\mathbf{M}_w/\rho - \mathbf{b}\right)} = \frac{\mathbf{a}\rho^2}{\mathbf{M}_w^2} + \frac{\rho\mathbf{R}\theta}{\mathbf{M}_w - \rho\mathbf{b}}$	$\frac{2\mathbf{a}\rho}{\mathbf{M}_w^2} + \frac{\mathbf{R}\theta}{\mathbf{M}_w} h \frac{\mathbf{b}\rho}{\mathbf{M}_w} - \left(\frac{\mathbf{R}\theta\left(\frac{\mathbf{b}}{\mathbf{M}_w^2} + 1\right)}{\mathbf{M}_w - \mathbf{b}\rho}\right)$
Redlich-Kwong's $\mathbf{p} = \frac{\rho\mathbf{R}\theta}{\mathbf{M}_w - \rho\mathbf{b}} - \frac{\mathbf{a}\rho^2}{\sqrt{\theta}\mathbf{M}_w(\mathbf{M}_w + \mathbf{b}\rho)}$	$\frac{-\mathbf{R}\theta}{\mathbf{M}_w}\frac{1}{\left(1 - \frac{\mathbf{b}}{\mathbf{M}_w}\rho\right)} + \frac{\mathbf{a}}{\mathbf{M}_w^2\sqrt{\theta}}\left(\left[\rho^2\mathbf{M}_w \ln\left\lvert\frac{\mathbf{b}\rho}{\mathbf{M}_w} + 1\right\rvert - 2\rho\mathbf{M}_w + \frac{2\mathbf{M}_w^2}{\mathbf{b}}\ln\left\lvert\frac{\mathbf{b}\rho}{\mathbf{M}_w} + 1\right\rvert - \frac{\mathbf{M}_w\rho}{\mathbf{b}} + \frac{\mathbf{M}_w^2}{\mathbf{b}^2}\ln\left\lvert\frac{\mathbf{b}}{\mathbf{M}_w} + 1\right\rvert\right]\right) + \frac{\mathbf{R}\theta}{\mathbf{M}_w\left(1 - \frac{\mathbf{b}}{\mathbf{M}_w}\rho\right)} - \frac{\mathbf{a}\rho}{\mathbf{M}_w^2\sqrt{\theta}\left(1 + \frac{\mathbf{b}}{\mathbf{M}_w}\rho\right)}$

Beattie-Bridgman's $\frac{R\theta\left(1-\frac{c\rho}{M_w\theta^3}\right)\rho^2}{M_w^{\;2}}\left(\frac{M_w}{\rho}+B_0\left(1-b\rho/M_w\right)\right)+\frac{aA_0\rho^3}{M_w^{\;3}}$	$\frac{R\theta}{M_w}\left(\ln\rho-\frac{2c\rho}{M_w^{\;2}\theta^2}\right)-\frac{B_0R\theta\left(2\rho+\frac{3c\rho^2}{2M_w\theta^3}\right)}{M_w^{\;2}}$ $-\frac{2B_0bR\theta\rho-\frac{3c\rho^3}{3M_w\theta^3}}{M_w^{\;2}}+\frac{aA_0R\theta\left(\frac{5}{3}\rho^3-\frac{6c\rho^5}{5M_w\theta^3}\right)}{M_w^{\;2}}$ $+\frac{R\rho^2}{2M_w}+\frac{B_0R\rho^3}{3M_w^{\;2}}-\frac{Rb\rho^4}{4M_w^{\;3}}$
Virial $pV_m=R\theta\left(1+\frac{B}{V_m}+\frac{C}{V_m^2}+\frac{D}{V_m^3}+\ldots\right)$	$\frac{R\theta}{M_w}\left(1+\ln\rho+\frac{3B}{M_w}+\frac{5C\rho^2}{2M_w^2}+\frac{7D\rho^3}{3M_w^3}\right)$

Inserting this into the earlier derived Vorticity-Bernoulli equation obtains,

$$\int_{s_1}^{s_2}\frac{\partial\underline{v}}{\partial t}\cdot ds+\int_{s_1}^{s_2}\nabla\frac{v^2}{2}\cdot ds-\int_{s_1}^{s_2}\nabla\times\underline{\omega}\cdot ds=\int_{s_1}^{s_2}\frac{1}{\rho}\cdot\left\{\left.\frac{\partial p}{\partial\rho}\right|_\theta+\left.\frac{\partial p}{\partial\theta}\right|_\rho\right\}ds$$

$$+\int_{s_1}^{s_2}\frac{1}{\rho}\frac{1}{A}\int\nabla\cdot\underline{\underline{T}}\cdot\underline{n}dA\cdot ds-\int_{s_1}^{s_2}\nabla\varphi\cdot ds \quad (5)$$

Writing with a velocity potential for time-dependence provides more familiar Bernouilli forms,

$$\underline{v}=\nabla\Omega$$

$$\int_{s_1}^{s_2}\nabla\cdot\left(\frac{\partial\Omega}{\partial t}+\frac{v^2}{2}+\varphi\right)\cdot ds=\int_{s_1}^{s_2}\frac{1}{\rho}\cdot\left\{\left.\frac{\partial p}{\partial\rho}\right|_\theta+\left.\frac{\partial p}{\partial\theta}\right|_\rho\right\}ds$$

$$+\int_{s_1}^{s_2}\frac{1}{\rho}\frac{1}{A}\int\nabla\cdot\underline{\underline{T}}\cdot\underline{n}dA\cdot ds$$

Then

$$\left(\frac{\partial\Omega}{\partial t}+\frac{v^2}{2}+\varphi \right)=\int_{s_1}^{s_2}\frac{1}{\rho}\cdot\left\{\frac{\partial p}{\partial\rho}\Big|_{\theta}+\frac{\partial p}{\partial\theta}\Big|_{\rho}\right\}ds+\int_{s_1}^{s_2}\frac{1}{\rho}\frac{1}{A}\int \nabla\cdot\underline{\underline{T}}\cdot\underline{n}dA\cdot ds+C(t)$$

$$\left(\frac{d}{dt}\int_{s_1}^{s_2} A\cdot\overline{V}\cdot ds+\frac{v^2}{2}+\varphi \right)=\int_{s_1}^{s_2}\frac{1}{\rho}\cdot\left\{\frac{\partial p}{\partial\rho}\Big|_{\theta}+\frac{\partial p}{\partial\theta}\Big|_{\rho}\right\}ds+\int_{s_1}^{s_2}\frac{1}{\rho}\frac{1}{A}\int \nabla\cdot\underline{\underline{T}}\cdot\underline{n}dA\cdot ds$$

The volume discharge rate may be written as an area average of the velocity tangent to the streamline across the stream tube cross-section.

$$Q=\int v\cdot dA=A\overline{\overline{V}} \qquad \overline{\overline{V}}=\frac{1}{A}\int v\cdot dA$$

$$\left(\frac{d}{dt}\int_{s_1}^{s_2} Q\cdot ds+\frac{v^2}{2}+\varphi \right)=\int_{s_1}^{s_2}\frac{1}{\rho}\cdot\left\{\frac{\partial p}{\partial\rho}\Big|_{\theta}+\frac{\partial p}{\partial\theta}\Big|_{\rho}\right\}ds+\int_{s_1}^{s_2}\frac{1}{\rho}\frac{1}{A}\int \nabla\cdot\underline{\underline{T}}\cdot\underline{n}dA\cdot d$$

Next consider a single chemical reaction,

$$\left(\frac{\partial\Omega}{\partial t}+\frac{v^2}{2}+\varphi \right)=\int_{s_1}^{s_2}\frac{1}{\rho}\cdot\left\{\frac{\partial p}{\partial\rho}\Big|_{\theta,\xi}d\rho+\frac{\partial p}{\partial\theta}\Big|_{\rho,\varsigma}d\theta+\frac{\partial p}{\partial\xi}\Big|_{\rho,\theta}d\xi\right\}ds+\int_{s_1}^{s_2}\frac{1}{\rho}\frac{1}{A}\int \nabla\cdot\underline{\underline{T}}\cdot\underline{n}dA\cdot ds+C(t)$$

Taking the differential of pressure,

$$dp=\left(\frac{\partial p}{\partial\rho}\dot{\rho}+\frac{\partial p}{\partial\theta}\dot{\theta}+\frac{\partial p}{\partial\xi}\dot{\xi}\right)d$$

Then using the balance of mass and substituting for $\dot{\rho}$

$$\mathbf{dp} = \left(-\rho \frac{\partial \mathbf{p}}{\partial \rho} \nabla \cdot \underline{\mathbf{v}} + \frac{\partial \mathbf{p}}{\partial \theta} \dot{\theta} + \frac{\partial \mathbf{p}}{\partial \xi} \dot{\xi} \right) \mathbf{dt}$$

$$\mathbf{p} = \int \mathbf{dp} = \int \left(-\rho \frac{\partial \mathbf{p}}{\partial \rho} \nabla \cdot \underline{\mathbf{v}} + \frac{\partial \mathbf{p}}{\partial \theta} \dot{\theta} + \frac{\partial \mathbf{p}}{\partial \xi} \dot{\xi} \right) \mathbf{dt} + \mathbf{cst}$$

$$\nabla \mathbf{p} = \nabla \left(-\rho \frac{\partial \mathbf{p}}{\partial \rho} \nabla \cdot \underline{\mathbf{v}} + \frac{\partial \mathbf{p}}{\partial \theta} \dot{\theta} + \frac{\partial \mathbf{p}}{\partial \xi} \dot{\xi} \right) \mathbf{dt} \quad ;$$

$$\nabla \mathbf{p} = \int \left(-\nabla \rho \frac{\partial \mathbf{p}}{\partial \rho} \cdot \nabla \cdot \underline{\mathbf{v}} - \rho \frac{\partial \mathbf{p}}{\partial \rho} \nabla^2 \underline{\mathbf{v}} + \nabla \left(\frac{\partial \mathbf{p}}{\partial \theta} \dot{\theta} \right) + \nabla \left(\frac{\partial \mathbf{p}}{\partial \xi} \dot{\xi} \right) \right) \mathbf{dt}$$

Further expansion of terms can be made if spatial homogeneity assumptions cannot be made using, for example,

$$\nabla (\mathbf{f} \nabla \mathbf{g}) = \mathbf{f} \cdot \nabla (\nabla \mathbf{g}) + (\nabla \mathbf{f}) \cdot (\nabla \cdot \mathbf{g}) = \mathbf{f} \nabla^2 \mathbf{g} + (\nabla \mathbf{f}) \cdot (\nabla \mathbf{g})$$

The above derivation has focused on the momentum and mass balances and now the local field equation for the internal energy in terms of temperature is provided and explicitly embedded into the integrated linear momentum equation that forms the Vorticicty-Bernouilli equation with well-known energy interpretation of its terms. The independent internal energy equation is written as the total differential with respect to time of the temperature being balanced by terms representing the viscous dissipation, the thermal transport and the chemical reaction progress. Refer to the Appendix A summary of the chemical reaction supply and the link to stoichiometry.

$$\dot{\theta} = -\frac{\theta_0 \beta_\theta}{\rho} \frac{\partial \mathbf{p}}{\partial \mathbf{t}} + \frac{1}{\rho \mathbf{c}_p} \mathbf{tr} \underline{\underline{\mathbf{TL}}} + \frac{1}{\rho \mathbf{c}_p} \nabla \cdot \underline{\mathbf{h}} + \frac{\mathbf{r}}{\mathbf{c}_p} - \frac{\rho \varsigma \overset{+}{\dot{\xi}}}{\mathbf{c}_p} \tag{6}$$

A Bernoulli equation obtained for the true irrotational, or no vorticity case,

$$\frac{\partial\Omega}{\partial t}+\frac{\mathbf{v}^2}{2}+$$
$$\frac{1}{\rho}\left[-\rho\frac{\partial \mathbf{p}}{\partial\rho}\nabla\cdot\underline{\mathbf{v}}+\frac{\partial \mathbf{p}}{\partial\rho}\left(-\frac{\theta_0\beta_\theta}{\rho}\frac{\partial \mathbf{p}}{\partial t}+\frac{1}{\rho \mathbf{c}_p}\mathbf{tr}\underline{\underline{\mathbf{TL}}}+\frac{1}{\rho \mathbf{c}_p}\nabla\cdot\underline{\mathbf{h}}+\frac{\mathbf{r}}{\mathbf{c}_p}-\frac{\rho\varsigma\overset{+}{\xi}}{\mathbf{c}_p}\right)+\frac{\partial \mathbf{p}}{\partial\xi}\dot{\xi}\right]\mathbf{dt}+\mathbf{C(t)} \tag{7}$$
$$+\overline{\Phi}_{\mathbf{LMVIS}}+\mathbf{d}\varphi=0$$

Here it appears useful to define terms expressible for solvable Poiseuile flow systems [7],

$$\overline{\Phi}_{\mathbf{LMVIS}}=-\frac{1}{\rho}\frac{1}{\mathbf{A}}\int\nabla\cdot\underline{\underline{\mathbf{T}}}\cdot\underline{\mathbf{n}}\mathbf{dAds}\qquad\Phi_{\mathbf{TVIS}}\equiv\frac{1}{\rho\mathbf{c}_p}\mathbf{tr}\,\underline{\underline{\mathbf{TL}}}$$

where $\mathbf{c}^2=\dfrac{\partial \mathbf{p}}{\partial\rho}$, and $\underline{\underline{\mathbf{L}}}=\nabla\underline{\mathbf{v}}$.

One may specify a thermal conductor with memory such as by a constitutive relation,

$$\mathbf{h}=-\frac{\mathbf{k}_\theta}{\tau_\mathbf{c}}\int e^{\frac{\Delta\theta}{\tau_\mathbf{c}}}\,\theta_{,\mathbf{x}}\,\mathbf{d}\theta_\mathbf{p}$$

The memory-less Fourier's Law where the heat flux is proportional to the temperature gradient $\underline{\mathbf{h}}=-\nabla(\mathbf{k}_\theta\theta)$ has attracted much attention in engineering and physics instruction. For homogeneous materials, obtain $\underline{\mathbf{h}}=-\mathbf{k}_\theta\nabla\theta$.

Next, consider taking into account end boundary motion by the rule of Leibnitz for differentiating an integral. This enables the acceleration term to be rewritten as follows.,

$$\frac{\mathbf{d}}{\mathbf{dt}}\int_{s_1}^{s_2}\underline{\mathbf{v}}\cdot\mathbf{ds}-\int_{s_1}^{s_2}\frac{\partial \mathbf{s}_2}{\partial t}\underline{\mathbf{v}}\cdot\mathbf{ds}+\int_{s_1}^{s_2}\frac{\partial \mathbf{s}_1}{\partial t}\underline{\mathbf{v}}\cdot\mathbf{ds}=\int_{s_1}^{s_2}\frac{\partial \underline{\underline{\mathbf{v}}}}{\partial t}\cdot\mathbf{ds} \tag{8}$$

$$\int_{s_1}^{s_2}\frac{\mathbf{d}\overline{\mathbf{V}}}{\mathbf{dt}}\cdot\mathbf{ds}-\int_{s_1}^{s_2}\frac{\partial \mathbf{s}_2}{\partial t}\underline{\mathbf{v}}\cdot\mathbf{ds}+\int_{s_1}^{s_2}\frac{\partial \mathbf{s}_1}{\partial t}\underline{\mathbf{v}}\cdot\mathbf{ds}=\int_{s_1}^{s_2}\frac{\partial \underline{\underline{\mathbf{v}}}}{\partial t}\cdot\mathbf{ds}\qquad\overline{\mathbf{V}}\equiv\frac{1}{\mathbf{T}}\int_{t_1}^{t_2}\underline{\mathbf{v}}\cdot\mathbf{dt} \tag{9}$$

This obtains the following forms for the Bernouilli equation,

$$\frac{\mathbf{d\overline{V}}}{\mathbf{dt}}-\frac{\partial \mathbf{s}_2}{\partial \mathbf{t}}\cdot\underline{\mathbf{v}}+\frac{\partial \mathbf{s}_1}{\partial \mathbf{t}}\cdot\underline{\mathbf{v}}+\frac{\mathbf{v}^2}{2}+\varphi=\int_{s_1}^{s_2}\frac{1}{\rho}\cdot\left\{\frac{\partial \mathbf{p}}{\partial \rho}\bigg|_{\theta,\xi}\mathbf{d}\rho+\frac{\partial \mathbf{p}}{\partial \theta}\bigg|_{\rho,\varsigma}\mathbf{d}\theta+\frac{\partial \mathbf{p}}{\partial \xi}\bigg|_{\rho,\theta}\mathbf{d}\xi\right\}\mathbf{ds}$$
$$+\int_{s_1}^{s_2}\frac{1}{\rho}\frac{1}{\mathbf{A}}\int\nabla\cdot\underline{\underline{\mathbf{T}}}\cdot\underline{\mathbf{n}}\mathbf{dA}\cdot\mathbf{ds}+ \qquad (10)$$

$$\frac{\mathbf{d\overline{V}}}{\mathbf{dt}}-\frac{\partial \mathbf{s}_2}{\partial \mathbf{t}}\cdot\underline{\mathbf{v}}+\frac{\partial \mathbf{s}_1}{\partial}\cdot\underline{\mathbf{v}}+\frac{\mathbf{v}^2}{2}+$$
$$\frac{1}{\rho}\left[-\rho\frac{\partial \mathbf{p}}{\partial \rho}\nabla\cdot\underline{\mathbf{v}}+\frac{\partial \mathbf{p}}{\partial \rho}\left(-\frac{\theta_0\beta_\theta}{\rho}\frac{\partial \mathbf{p}}{\partial \mathbf{t}}+\frac{1}{\rho \mathbf{c}_p}\mathbf{tr}\underline{\underline{\mathbf{TL}}}+\frac{1}{\rho \mathbf{c}_p}\nabla\cdot\underline{\mathbf{h}}+\frac{\mathbf{r}}{\mathbf{c}_p}-\frac{\rho\varsigma\overset{+}{\xi}}{\mathbf{c}_p}\right)+\frac{\partial \mathbf{p}}{\partial \xi}\dot{\xi}\right]\mathbf{dt}$$
$$+\overline{\Phi}_{\mathbf{LMVIS}}+\mathbf{d}\varphi=0 \qquad (11)$$

This averaging approach offers an alternative that may be useful for periodic time-dependent systems. The time averages are equivalent to ensemble average for an ergodic process. The former equation would be most useful when thermodynamic equation of state estimates are available while the latter relies on the internal energy equation's parameters and constitutive relations.

The approximate case when the vorticity is zero or the flow is irrotational with constant density; in addition to the flow being steady and without thermal, viscous, or chemical reaction dissipation, but in a vertical gravity body force field upon integrating along a streamline with fixed boundaries (i.e., $s_{1,}$ s_2 are constant) gives the commonly applied form of the Bernoulli's equation.

$$\left[\frac{\mathbf{p}}{\rho}+\frac{1}{2}\mathbf{v}^2+\mathbf{gz}\right]_{s_1}^{s_2}=0$$

It expresses that the total energy (pressure plus gravitational plus kinetic energy) is constant. Refer to Appendix B on reincorporation of dissipation and loss including boundary sources of vorticity for the severe assumptions used in its derivation for engineering application.. Generalizing from this equation can be achieved using the energy or enthalpy balances.

In thermodynamic contexts the enthalpy potential **H** is typically introduced for homogeneous, static, constant pressure thermodynamic systems without dissipation. This may now be used to rewrite the energy and vorticity-Bernoulli equations. Let $\mathbf{H} = \varepsilon + \mathbf{p}\nu, \quad \dot{\mathbf{H}} = \dot{\varepsilon} + \mathbf{p}\dot{\nu} + \dot{\mathbf{p}}\nu$

The internal energy equation incorporating thermal conduction $\underline{\mathbf{h}}$, radiation $\mathbf{r}$, viscous dissipation $\mathbf{tr}\underline{\underline{\mathbf{SD}}}$, and chemical reaction heat ς writes,

$$\rho\dot{\varepsilon} = \mathbf{tr}\underline{\underline{\mathbf{SD}}} + \nabla \cdot \underline{\mathbf{h}} + \rho\mathbf{r} - \rho\varsigma\dot{\xi} \tag{12}$$

Steps of the derivation include,

$$\mathbf{tr}\underline{\underline{\mathbf{SD}}} = \mathbf{tr}[(-\mathbf{p}\underline{\underline{\mathbf{I}}} + \underline{\underline{\mathbf{T}}})\mathbf{D}] = -\mathbf{p}\nabla \cdot \underline{\mathbf{v}} + \mathbf{tr}\underline{\underline{\mathbf{TD}}}$$

$$\dot{\mathbf{H}} - \mathbf{p}\dot{\nu} - \dot{\mathbf{p}}\nu = \dot{\varepsilon} = \frac{-\mathbf{p}}{\rho}\nabla \cdot \underline{\mathbf{v}} + \frac{1}{\rho}\mathbf{tr}\underline{\underline{\mathbf{TD}}} + \frac{1}{\rho}\nabla \cdot \underline{\mathbf{h}} + \mathbf{r} - \varsigma\dot{\xi}$$

where $\dot{\nu} = -\frac{1}{\rho^2}\dot{\rho} = \frac{1}{\rho}\nabla \cdot \underline{\mathbf{v}} = \upsilon\nabla \cdot \underline{\mathbf{v}}$ The specific volume, or reciprocal density is denoted by ν. The enthalpy equation with dissipation becomes,

$$\dot{\mathbf{H}} = \dot{\mathbf{p}}\nu + \nu \cdot \mathbf{tr}\underline{\underline{\mathbf{TD}}} + \frac{1}{\rho}\nabla \cdot \underline{\mathbf{h}} + \mathbf{r} - \varsigma\dot{\xi} \tag{13}$$

Again the special case of constant pressure, systems with negligible dissipation such as inviscid, non-heat conducting, non-heat radiating, and non-reacting material flows has exploited this potential function frequently.

Vorticity and Circulation

The purpose of this section is to re-emphasize the equation development within a framework that explicitly addresses both vorticity and circulation and applies the thermodynamic equations previously derived.[14,15] The vorticity defined as $\underline{\omega} = \nabla \times \underline{\mathbf{v}}$ represents twice the local rotation of the fluid element. The linear momentum equation of motion in local form is repeated below.

$$\frac{\partial \underline{\mathbf{v}}}{\partial \mathbf{t}} - \underline{\mathbf{v}} \times \underline{\omega} = -\frac{1}{\rho} \nabla \mathbf{p} + \nabla \cdot \underline{\mathbf{T}}$$

This is written in a modified form by taking its curl. Here a Newtonian stress tensor is assumed for illustration,

$$\underline{\mathbf{T}} = \mu \nabla \underline{\mathbf{v}} + \left(\frac{\mu}{3} + \mu_v \right) \nabla (\nabla \cdot \underline{\mathbf{v}})$$

obtains $\frac{\partial \underline{\omega}}{\partial \mathbf{t}} - \nabla \times (\underline{\mathbf{v}} \times \underline{\omega}) = -\frac{1}{\rho^2} \nabla \rho \times \nabla \mathbf{p} + \nu \nabla^2 \underline{\omega}$ with kinematic viscosity $\nu = \frac{\mu}{\rho}$

The circulation is defined as an integral around the circuit which may be related to the curl in the normal direction to a surface element. [16]

$$\mathbf{C} = \oint \underline{\mathbf{v}} \cdot \mathbf{d}\underline{\mathbf{l}} = \int \underline{\omega} \cdot \mathbf{d}\underline{\mathbf{s}}$$

such that $\underline{\mathbf{n}} \cdot \nabla \times \underline{\mathbf{v}} = \frac{1}{\delta \mathbf{S}} \oint \underline{\mathbf{v}} \cdot \mathbf{d}\underline{\mathbf{l}}$. Using

$$\nabla \times (\underline{\mathbf{u}} \times \underline{\omega}) = \underline{\omega} \cdot \nabla \underline{\mathbf{v}} - (\underline{\mathbf{v}} \cdot \nabla) \underline{\omega} - \underline{\omega} \nabla \cdot \underline{\mathbf{v}} + \underline{\mathbf{v}} \nabla \cdot \underline{\omega}$$

the dynamic vorticity equation may be written as

$$\frac{\partial \underline{\omega}}{\partial \mathbf{t}} + (\underline{\mathbf{u}} \cdot \nabla) \underline{\omega} = \underline{\omega} \cdot \nabla \underline{\mathbf{v}} - \underline{\omega} \nabla \cdot \underline{\mathbf{v}} = -\frac{1}{\rho^2} \nabla \rho \times \nabla \mathbf{p} + \nu \nabla^2 \underline{\omega}$$

Now the earlier derivation for the gradient of pressure and for incorporation of the energy equation may proceed here in a parallel fashion for the vorticity equation.

That is, $\nabla \mathbf{p} = \int \left(-\nabla \rho \frac{\partial \mathbf{p}}{\partial \rho} \cdot \nabla \cdot \underline{\mathbf{v}} - \rho \frac{\partial \mathbf{p}}{\partial \rho} \nabla^2 \underline{\mathbf{v}} + \nabla \left(\frac{\partial \mathbf{p}}{\partial \theta} \dot{\theta} \right) + \nabla \left(\frac{\partial \mathbf{p}}{\partial \xi} \dot{\xi} \right) \right) \mathbf{dt}$ and

$\dot{\theta} = -\frac{\theta_0 \beta_\theta}{\rho} \frac{\partial \mathbf{p}}{\partial \mathbf{t}} + \frac{1}{\rho \mathbf{c}_p} \mathbf{tr} \underline{\underline{\mathbf{TL}}} + \frac{1}{\rho \mathbf{c}_p} \nabla \cdot \underline{\mathbf{h}} + \frac{\mathbf{r}}{\mathbf{c}_p} - \frac{\rho \varsigma \overset{+}{\xi}}{\mathbf{c}_p}$ Particular equations of state may now be inserted directly into the analysis of vorticity dynamics. For the special case when pressure is only a function of density, $\frac{1}{\rho^2}\nabla\rho\times\nabla\mathbf{p} = \frac{1}{\rho^2}\nabla\rho\times\nabla\rho\frac{\mathbf{dp}}{\mathbf{d}\rho} \Rightarrow 0$ and the vorticity equation greatly simplifies.

Conclusion

Extensions of the Vorticity-Bernouili equations enables a better understanding of assumptions of simpler derivations of the equation aiming for a particular system that usually involves equilibrium or steady state conditions. Furthermore, more general applications are more apparent when including equation of state, end boundary motions, or other phenomenology such as chemical reaction, or other constitutive relation assumptions, such as the dissipative viscous or thermal memory transfers. These reformulations guide those previously given aiming for broader perspectives and knowledge with understanding in application to experience. Further investigations coupling thermodynamics and fluid flows with fluctuations about a mean value also provide many useful applications.

Appendix A

Consider chemical reactions $\sum_{\alpha=1}^{k} \mathbf{S}_\alpha \mathbf{C}_\alpha = 0$ involving constituents $C_{\alpha,}$ $\alpha = 1,$

where the reactants $\alpha = 1, k_R,$ Reaction products are $\alpha = k_R+1, k.$

The signed stoichiometric coefficients $\mathbf{S}_\alpha$; with Sign $[\mathbf{S}_\alpha] < 0$ for reactants; the Sign $[\mathbf{S}_\alpha] > 0$ for products; and $\mathbf{S}_\alpha = 0$ for an inert constituent. For the case when diffusion can be neglected as applicable to many liquid mixtures at ordinary temperatures and pressures a degree of advancement,

or reaction progress variables ξ, $\omega - \omega^e = \sum_{\alpha=1}^{k} \mathbf{S}_\alpha \xi$ may be defined [8,9].

Chemical composition is described by $[\omega_\alpha]$ or gmol/mass; the chemical kinetic equations set using compositions is usually larger than one using

reaction progress or reaction velocity for kinetic analysis: $\xi' = \overset{+}{\xi}$.

Appendix B

The engineering analysis proceeds using assumptions to derive and apply [10-13], $\left[\frac{\mathbf{p}}{\gamma}+\alpha\frac{\mathbf{v}^2}{2\mathbf{g}}+\mathbf{z}\right]_1^2+\mathbf{h}_L+\mathbf{h}_s=0$ Source terms are added as needed. The pressure head term, $\mathbf{p}/\gamma$, includes material properties for the fluid, such as the specific weight, $\gamma=\rho\mathbf{g}$, the density, ρ, and the dynamic shear viscosity, μ. Illustrative fluid properties are given for approximately 70 $^{\circ}$F:

Fluid	Specific Weight (lbf/ft^3)	Density (lbm/ft^3)	Dynamic Viscosity (lbm/ft-s)
water	62.3	62.3	$6.56\text{x}10^{-4}$
gasoline	42.5	42.5	$1.96\text{x}10^{-4}$
engine oil	55.4	55.4	$5.8\text{x}10^{-1}$

The friction loss in a piping system affects the flow rate as estimated by, for example by a head loss term $\mathbf{h}_L$.

$\mathbf{h}_L=\mathbf{f}\frac{\mathbf{L}_p}{\mathbf{D}_h}\frac{\mathbf{v}^2}{2\mathbf{g}}+\left(\sum_i \mathbf{K}_i\right)\frac{\mathbf{v}^2}{2\mathbf{g}}$ For laminar flow the dimensionless Darcy friction factor $\mathbf{f}=\frac{64}{\mathbf{Re}}$, while for turbulent flow empirical data or correlations (Swamee-Jain) are applied, $\mathbf{f}=\frac{0.25}{\left[\mathbf{log}\left(\frac{\varepsilon/\mathbf{D}_h}{3.7}+\frac{5.74}{\mathbf{Re}_{0,9}}\right)\right]^2}$.

$\mathbf{Re}=\frac{\rho\mathbf{D}_h\mathbf{v}}{\eta}$ is a dimensionless Reynold's Number, D_h = the hydraulic diameter, L_p = the length of pipe, $\mathbf{K}_i$ represents the loss coefficient for the i-th minor loss component in the piping system, and $\varepsilon/\mathbf{D}_h$ is a relative roughness measure.

Simple Dimensional Analysis Applied to Poiseuille Flow

Consider a cylindrical pipe geometry of radius R for a steady isothermal system.

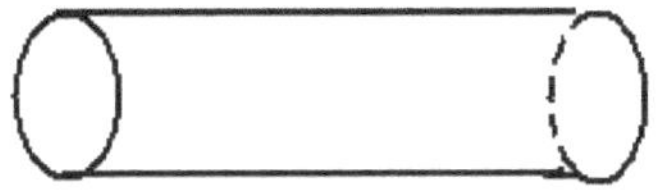

Let the volume flow rate be denoted by Q where the dimensions of the parameter Q are $[L^3T^{-1}]$. Assuming a dependence on the pressure gradient $\frac{\partial p}{\partial x}$ (here taken as along the axial coordinate z), the radius R, as well as the shear viscosity μ one may write using fundamental dimensions of length L, time T, and mass M to formulate by dimensional analysis the general dependence on the parameters of the flow.:

$$\frac{\partial p}{\partial z} \sim \left[\frac{\frac{ML}{T^2}\cdot\frac{1}{L^2}}{L}\right] = \left[\frac{M}{L^2T^2}\right]; \qquad \mu \sim \left[\frac{\tau}{\frac{\partial v}{\partial z}}\right] = \left[\frac{\frac{\frac{M}{LT^2}}{L}}{\frac{T}{L}}\right] = \left[\frac{M}{LT}\right]; \qquad R = [L]$$

Using the hypothesized function dependence,

$$Q \sim f\left(\frac{\partial p}{\partial z}, \mu, R\right) \rightarrow Q\left(\frac{M}{L^2T}^{2}\right)^{\alpha}\left(\frac{M}{LT}\right)^{\beta} L^{\gamma} \sim 1.$$

The exponents form the following simultaneous algebraic system of equations to be solved,

$$\begin{aligned}&\alpha + \beta = 0\\ &3 + (-2\alpha) + (-\beta) + \gamma = 0 \quad \text{obtains} \quad \alpha = -1, \quad \beta = 1, \quad \gamma = -4\\ &-1 + (-2\alpha) + (-\beta) = 0\end{aligned}$$

Therefore, $Q\left(\frac{\partial p}{\partial z}\right)^{-1}\mu^{1}R^{-4} \sim 1$ or rewriting for $Q \sim \frac{1}{\mu}\left(\frac{\partial p}{\partial z}\right)R^{4} \sim 1$

The results show that the volume flow rate is directly proportional to both the pressure gradient and radius to the fourth power, as well as inversely proportional to the shear viscosity. Next consider, including pipe roughness as measured by an average size of unevenness along the pipe surface length in the list of variables (parameters) for this problem denoted by [E]. Form a parameter set along with the density, average pipe radius [R] (or diameter D=2R), pressure gradient $\left[\frac{\mathbf{dp}}{\mathbf{dz}}\right]$, shear viscosity [μ] (or kinematic viscosity [ν]= $\left[\frac{\mu}{\rho}\right]$, and volume flow rate or velocity [v]. This results in six parameters with three fundamental units of mass, length, and time which implies that there are 6 - 3 = 3 independent dimensionless quantities from Buckingham's Pi-Theorem.[17,18] The maximal number of three have been typically identified for use as:

Reynold's Number = $\left[\frac{\mathbf{v}\cdot\mathbf{D}}{\nu}\right]$,

Relative roughness $\left[\varepsilon = \frac{\mathbf{E}}{\mathbf{D}}\right]$, and $\left[\frac{\frac{\mathbf{dp}}{\mathbf{dz}}\cdot\mathbf{D}}{\rho\mathbf{v}^2}\right]$.

Statistical analysis of experiments have provided engineering analysis formulas in functions such as $\left[\frac{\frac{\mathbf{dp}}{\mathbf{dz}}\cdot\mathbf{D}}{\rho\mathbf{v}^2}\right] = \mathbf{g}(\mathrm{Re},\varepsilon)$. Other measures of fractal surface geometry roughness may also be defined and useful for investigation.

References

1. Truesdell, C.A., *Rational Thermodynamics* (McGraw Hill, 1984).
2. Slattery, John C., *Momentum, Energy, and Mass Transfer in Continua* (R. Krieger Publishing Co, 1981).
3. Bird, R. W.E. Stewart, and N. Lightfoot, *Transport Phenomena* (Wiley, 1960).

4. Milne-Thompson, *Theoretical Hydrodynamics* (Macmillon, 1953).
5. Landau, L. and E.M. Lifshitz Fluid Mechanics (Addison Wesley, 1960).
6. Astarita, G. and M. Mackay, J. Rheol., 40,(3), (1996).
7. Synolakis, C. and H. Badeer, *Am. J. Phys.*, Vol. 57, No. 11 (1989).
8. Prigogine, I. and R. Defay, *Chemical Thermodynamics*, Translated by D.H. Everett (Longman, 1954).
9. Bowen, R.M., *Arch. Ratl. Mech. Anal.*, *24*, 370 (1967); *J. Chem.* Phys. *49*, 1625 (1968).
10. Cengel, Y. and R. Turner, *Fundamentals of Fluid-Thermal Sciences* (McGraw-Hill, 2006).
11. Moody, L.F., *Trans ASME*, 66, 671 (1944).
12. P.E. Swamee and A.K. Jain, *J. Hydraulics Div.*, *102* (HY5), 657, Am. Soc. Civil Engineers (1976).
13. "Multi-Semester Interwoven Project for Teaching Basic Core STEM [Science, Technology, Engineering, Mathematics], University of Massachusetts-Lowell NSF Project.
14. Truesdell, C., "The Kinematics of Vorticity," Indiana Univ. Press (1954).
15. Truesdell, C. *J. Rat. Mech Anal.*, *2* (1953) 173-217.
16. Snieder, R., *A Guided Tour of Math Methods for the Physical Sciences*, Cambridge (2004).
17. Buckingham, E., On Physically Similar Systems: Illustrations of the Use of Dimensional Equations, *Phys. Rev.*, *4*, 345-376 (1914).
18. Olsen, Harold Hanche, "Buckingham's pi-Theorem", TMA4195 Mathematical Modeling (2004).

Integration Notes:

Equation of State	$\int \frac{\mathbf{dp}}{\rho}$; $\mathbf{dp}(\rho,\theta)=\frac{\partial \mathbf{p}}{\partial \rho}\Big|_{\theta}\mathbf{d}\rho+\frac{\partial \mathbf{p}}{\partial \theta}\Big|_{\mathbf{p}}\mathbf{d}\theta$
Tait's $\mathbf{p}=\mathbf{A}\left[\left(\frac{\rho}{\rho_0}\right)^{\mathbf{z}}-1\right]$	$\int \frac{\mathbf{dp}(\rho)}{\rho}=\frac{\mathbf{A}}{\rho_0^{\mathbf{z}}}\frac{\mathbf{z}}{(\mathbf{z}-1)}\int \rho^{\mathbf{z}-1}\mathbf{d}\rho=$ $\frac{\mathbf{A}}{\rho_0^{\mathbf{z}}}\frac{\mathbf{z}}{(\mathbf{z}-1)}\frac{\rho^{\mathbf{z}}}{\mathbf{z}}=\frac{\mathbf{A}}{\rho_0^{\mathbf{z}}}\frac{\rho^{\mathbf{z}}}{(\mathbf{z}-1)}$
Van der Waal's $\mathbf{p}=\frac{\mathbf{a}\rho^2}{\mathbf{M}_{\mathbf{w}}^2}+\frac{\rho\mathbf{R}\theta}{\mathbf{M}_{\mathbf{w}}-\rho\mathbf{b}}$	$\mathbf{dp}=\left[\frac{\mathbf{2a}\rho}{\mathbf{M}_{\mathbf{w}}^2}+\frac{\mathbf{R}\theta}{\mathbf{M}_{\mathbf{w}}-\rho\mathbf{b}}+\rho\mathbf{R}\theta\frac{\mathbf{b}}{(\mathbf{M}_{\mathbf{w}}-\rho\mathbf{b})^2}\right]\mathbf{d}\rho+\frac{\rho\mathbf{R}}{\mathbf{M}_{\mathbf{w}}-\rho\mathbf{b}}\mathbf{d}\theta$ $\int\frac{\mathbf{dp}}{\rho}=\int\frac{\mathbf{2a}\cdot\mathbf{d}\rho}{\mathbf{M}_{\mathbf{w}}^2}+\int\frac{\mathbf{R}\theta\cdot\mathbf{d}\rho}{\rho(\mathbf{M}_{\mathbf{w}}-\rho\mathbf{b})}+\int\mathbf{R}\theta\frac{\mathbf{b}\cdot\mathbf{d}\rho}{(\mathbf{M}_{\mathbf{w}}-\rho\mathbf{b})^2}+\int\frac{\mathbf{R}\cdot\mathbf{d}\theta}{\mathbf{M}_{\mathbf{w}}-\rho\mathbf{b}}$ Since $\int\frac{\mathbf{R}\theta\cdot\mathbf{d}\rho}{\rho(\mathbf{M}_{\mathbf{w}}-\rho\mathbf{b})}+\int\mathbf{R}\theta\frac{\mathbf{b}\cdot\mathbf{d}\rho}{(\mathbf{M}_{\mathbf{w}}-\rho\mathbf{b})^2}$ $=\frac{-\mathbf{R}\theta}{\mathbf{b}}\int\frac{\mathbf{d}\rho}{\rho\left(\rho-\frac{\mathbf{M}_{\mathbf{w}}}{\mathbf{b}}\right)}+\frac{\mathbf{bR}\theta}{\mathbf{M}^2_{\mathbf{w}}}\int\frac{\mathbf{d}\rho}{\left(1-\rho\frac{\mathbf{b}}{\mathbf{M}_{\mathbf{w}}}\right)^2}$ $=\frac{\mathbf{R}\theta}{\mathbf{b}}\left[\frac{\mathbf{b}}{\mathbf{M}_{\mathbf{w}}}\mathbf{ln}\,\rho-\frac{\mathbf{b}}{\mathbf{M}_{\mathbf{w}}}\mathbf{ln}\left|\rho-\frac{\mathbf{M}_{\mathbf{w}}}{\mathbf{b}}\right|\right]+\frac{\mathbf{R}\theta}{\mathbf{M}_{\mathbf{w}}}\mathbf{ln}\left|1-\frac{\mathbf{b}}{\mathbf{M}_{\mathbf{w}}}\rho\right|$ $=\frac{\mathbf{R}\theta}{\mathbf{M}_{\mathbf{w}}}\left[\mathbf{ln}\left|\frac{\rho}{\rho-\frac{\mathbf{M}_{\mathbf{w}}}{\mathbf{b}}}\right|\right]+\frac{\mathbf{R}\theta}{\mathbf{M}_{\mathbf{w}}}\mathbf{ln}\left|1-\frac{\mathbf{b}}{\mathbf{M}_{\mathbf{w}}}\rho\right|=\frac{\mathbf{R}\theta}{\mathbf{M}_{\mathbf{w}}}\mathbf{ln}\frac{\mathbf{b}\rho}{\mathbf{M}_{\mathbf{w}}}$ $\int\frac{\mathbf{dp}}{\rho}=\frac{\mathbf{2a}\rho}{\mathbf{M}_{\mathbf{w}}^2}+\frac{\mathbf{R}\theta}{\mathbf{M}_{\mathbf{w}}}\mathbf{ln}\frac{\mathbf{b}\rho}{\mathbf{M}_{\mathbf{w}}}-\left(\frac{\mathbf{R}\theta\left(\frac{\mathbf{b}}{\mathbf{M}^2_{\mathbf{w}}}+1\right)}{\mathbf{M}_{\mathbf{w}}-\mathbf{b}\rho}\right)$

Redlich-Kwong's $p = \frac{\rho R\theta}{M_w - \rho b} - \frac{a\rho^2}{\sqrt{\theta}M_w(M_w + b\rho)}$ $p = \frac{R}{M_w}\frac{\rho\theta}{\left(1-\rho\frac{b}{M_w}\right)} - \frac{a}{M_w^2}\frac{\rho^2\theta^{-0.5}}{\left(1+\frac{b}{M_w}\rho\right)}$	$\frac{1}{\rho}\frac{\partial p}{\partial \rho}\Big\|_\theta = \frac{\partial}{\partial \rho}\left(\frac{R}{M_w}\frac{\theta}{\left(1-\rho\frac{b}{M_w}\right)}\right) - \frac{\partial}{\partial \rho}\left(\frac{a}{M_w^2}\frac{\rho\,\theta^{-0.5}}{\left(1+\frac{b}{M_w}\rho\right)}\right)$ $= \frac{-bR\theta}{M_w^2}\frac{1}{\left(1-\frac{b}{M_w}\rho\right)^2} + \frac{a}{M_w^2\sqrt{\theta}}\left(\frac{b\rho^2-\rho}{\left(1+\frac{b}{M_w}\rho\right)}\right)$ $\frac{1}{\rho}\frac{\partial p}{\partial \theta}\Big\|_\rho = \frac{\partial}{\partial \theta}\left(\frac{R\theta}{M_w\left(1-\frac{b}{M_w}\rho\right)}\right) - \frac{a}{M_w^2}\frac{\partial}{\partial \theta}\left(\frac{\rho\theta^{-0.5}}{\left(1+\frac{b}{M_w}\rho\right)}\right)$ $= \frac{R}{M_w\left(1-\frac{b}{M_w}\rho\right)} + \frac{a\rho}{2M_w^2\theta\sqrt{\theta}\left(1+\frac{b}{M_w}\rho\right)}$ $\int \frac{1}{\rho}\frac{\partial p}{\partial \rho}\Big\|_\theta d\rho =$ $\frac{-bR\theta}{M_w^2}\int\frac{d\rho}{\left(1-\frac{b}{M_w}\rho\right)^2} + \frac{a}{M_w^2\sqrt{\theta}}\int\frac{b\rho^2-\rho}{\left(1+\frac{b}{M_w}\rho\right)}\cdot d\rho$ $= \frac{-R\theta}{M_w}\frac{1}{\left(1-\frac{b}{M_w}\rho\right)} + \frac{a}{M_w^2\sqrt{\theta}}\left(\left[\begin{array}{l}\rho^2 M_w \ln\left\|\frac{b\rho}{M_w}+1\right\| \\ -2\rho M_w + \frac{2M_w^2}{b}\ln\left\|\frac{b\rho}{M_w}+1\right\| \\ -\frac{M_w\rho}{b} + \frac{M_w^2}{b^2}\ln\left\|\frac{b}{M_w}+1\right\|\end{array}\right]\right)$ $\int \frac{1}{\rho}\frac{\partial p}{\partial \theta}\Big\|_\rho d\theta =$ $= \frac{R\theta}{M_w\left(1-\frac{b}{M_w}\rho\right)} - \frac{a\rho}{M_w^2\sqrt{\theta}\left(1+\frac{b}{M_w}\rho\right)}$

	$\int \frac{dp}{\rho} = \frac{-R\theta}{M_w}\frac{1}{\left(1-\frac{b}{M_w}\rho\right)} + \frac{a}{M_w^2\sqrt{\theta}}\left(\left[\begin{array}{l}\rho^2 M_w \ln\left\lvert\frac{b\rho}{M_w}+1\right\rvert \\ -2\rho M_w + \frac{2M_w^2}{b}\ln\left\lvert\frac{b\rho}{M_w}+1\right\rvert \\ -\frac{M_w\rho}{b}+\frac{M_w^2}{b^2}\ln\left\lvert\frac{b}{M_w}+1\right\rvert\end{array}\right]\right) + \frac{R\theta}{M_w\left(1-\frac{b}{M_w}\rho\right)} - \frac{a\rho}{M_w^2\sqrt{\theta}\left(1+\frac{b}{M_w}\rho\right)}$
Beattie-Bridgman's $p = \frac{R\theta\left(1-\frac{c\rho}{M\theta^3}\right)\rho^2}{M_w{}^2}\left(\frac{M_w}{\rho}+B_0\left(1-b\rho/M_w\right)\right)+\frac{aA_0\rho^3}{M_w{}^3}$ $p = \frac{\rho R\theta}{M_w}+\frac{RC\rho^2}{M_w^2\theta^2}+\frac{B_0R\theta\rho^2}{M_w^2}-\frac{B_0R\theta b\rho^2}{M_w^3}+\frac{B_0RC\rho^3}{M_w^3\theta^2}-\frac{B_0bRC\rho^3}{M_w^4\theta^2}+\frac{aA_0\rho^3}{M_w{}^3}$	$\int \frac{1}{\rho}\frac{\partial p}{\partial \rho}\Big\rvert_\theta = \frac{R\theta}{M_w}\ln\rho + \frac{2Rc\rho}{M_w^2\theta^2}+\frac{2B_0R\theta\rho}{M_w^2}-\frac{2B_0bR\theta\rho}{M_w^3}+\frac{3}{2}\frac{B_0Rc\rho^2}{M_w^3\theta^2}-\frac{3B_0Rcb\rho}{2M_w^4\theta^2}+\frac{3aA_0\rho^2}{2M_w^3}$ $\int \frac{1}{\rho}\frac{\partial p}{\partial \theta}\Big\rvert_\rho = \frac{R\theta}{M_w}+\frac{c\rho}{M_w^2\theta^2}+\frac{B_0R\theta\rho}{M_w^2}-\frac{B_0bR\theta\rho}{M_w^3}+\frac{B_0Rc\rho^2}{M_w^3\theta^2}-\frac{B_0Rc\rho^2}{M_w^4\theta^2}$ $\int \frac{1}{\rho}dp = \frac{R\theta}{M_w}\left(\ln\rho+\frac{2c\rho}{M_w\theta^3}\right)+\frac{B_0R\theta\left(2\rho+\frac{3c\rho^2}{2M_w\theta^3}\right)}{M_w^2}-\frac{2B_0bR\theta\rho}{M_w^3}+\frac{3}{2}\frac{B_0RC\rho^2}{M_w^3\theta^2}-\frac{3B_0RCb\rho}{2M_w^4\theta^2}+\frac{3aA_0\rho^2}{2M_w^3}+\frac{R\theta}{M_w}+\frac{C\rho}{M_w^2\theta^2}+\frac{B_0R\theta\rho}{M_w^2}-\frac{B_0bR\theta\rho}{M_w^3}+\frac{B_0RC\rho^2}{M_w^3\theta^2}-\frac{B_0RC\rho^2}{M_w^4\theta^2}$

<table>
<tr>
<td>Virial
$pV_m = R\theta\left(1+\frac{B}{V_m}+\frac{C}{V_m^2}+\frac{D}{V_m^3}+...\right)$
$V_m = V/n$
$p=\frac{R\theta\rho}{M_w}\left(1+\frac{B\rho}{M_w}+\frac{C\rho^2}{M_w^2}+\frac{D\rho^3}{M_w^3}+...\right)$
$V_m = V/n = \frac{M}{\rho n} = \frac{M_w}{\rho}$ $\quad \rho V = M$
$n=\frac{M}{M_w}$</td>
<td>$\frac{\partial p}{\partial\rho}\Big|_\theta = \frac{R\theta}{M_w}\left(1+\frac{2\rho B}{M_w}+\frac{3\rho^2C}{M_w^2}+\frac{4\rho^3D}{M_w^3}+.\right)$
$\frac{\partial p}{\partial\theta}\Big|_\rho = \frac{R\rho}{M_w}\left(1+\frac{\rho B}{M_w}+\frac{\rho^2 C}{M_w^2}+\frac{\rho^3 D}{M_w^3}+.\right)$
$\int\frac{dp}{\rho} = \frac{R\theta}{M_w}\int\left(\rho^{-1}+\frac{2B}{M_w}+\frac{3C\rho}{M_w^2}+\frac{4D\rho^2}{M_w^3}\right)d\rho$
$+\frac{R}{M_w}\int\left(1+\frac{B\rho}{M_w}+\frac{C\rho^2}{M_w^2}+\frac{D\rho^3}{M_w^3}\right)d\theta$
$=\frac{R\theta}{M_w}\left(1+\ln\rho+\frac{3B}{M_w}+\frac{5C\rho^2}{2M_w^2}+\frac{7D\rho^3}{3M_w^3}\right)$</td>
</tr>
</table>

Using calculus integration techniques of substitution, of newly defined variables, of integration by parts, or of partial fraction expansion the following useful formulas may be derived and applied in the hydrothermodynamic analysis:

$$\int\frac{x\cdot dx}{(ax+b)} = \frac{x}{a}-\frac{b}{a^2}\ln|ax+b|$$

$$\int\frac{x^2\cdot dx}{(ax+b)} = \frac{x^2}{a}\ln|ax+b|-2\left[\frac{x}{a}-\frac{b}{a^2}\ln|ax+b|\right]$$

$$\int\frac{dx}{(ax+b)^2} = -\frac{1}{a(ax+b)}$$

$$\int\frac{x\cdot dx}{(ax+b)^2} = \frac{b}{a^2(ax+b)}+\frac{1}{a^2}\ln|ax+b|$$

$$\int\frac{dx}{x(x+a)} = \frac{\ln x}{a}-\frac{1}{a}\ln|x+a|$$

XIV

Helical Flow Of A Third Grade Fluid

Analysis Summary

Using cylindrical coordinates $(\mathbf{r}, \theta, \mathbf{z})$ the equations of motion in differential form with velocity vector components $\underline{\mathbf{v}} = (\mathbf{u}_r, \mathbf{u}_\theta, \mathbf{u}_z)$ are written:

$$\nabla \cdot \underline{\mathbf{v}} = 0 \qquad \text{(Balance of Mass – Incompressible)} \quad (1)$$

$$\nabla \cdot \underline{\underline{\mathbf{T}}} + \rho \underline{\mathbf{f}} = \rho \frac{\mathbf{D}\underline{\mathbf{v}}}{\mathbf{Dt}} \qquad \text{(Balance of Linear Momentum)} \quad (2)$$

$$\rho \theta \frac{\mathbf{D}\eta}{\mathbf{Dt}} = \mathbf{tr} \underline{\underline{\mathbf{T}}} \cdot \underline{\underline{\mathbf{L}}} - \nabla \cdot \underline{\mathbf{h}} \qquad \text{(Balance of Entropy/Energy)}$$

where η denotes the entropy. For internal energy a function of temperature Θ and pressure p, the entropy equation may be written in the familiar form at constant pressure,

$$\rho \mathbf{c}_p \frac{\mathbf{D}\Theta}{\mathbf{Dt}} = \mathbf{tr} \underline{\underline{\mathbf{S}}} \cdot \underline{\underline{\mathbf{L}}} - \nabla \cdot \underline{\mathbf{h}} \qquad (3)$$

$\underline{\mathbf{h}}$ is the heat flux vector and $\underline{\underline{\mathbf{L}}}$ represents the velocity gradient.

These equations of motion are combined with constitutive equation assumptions for the stress tensor and heat flux vector for application to real materials. The total stress $\underline{\underline{T}}$ for an incompressible homogeneous third grade simple fluid as the extra stress $\underline{\underline{S}}$ is written as follows:

$$\underline{\underline{T}} = -p\underline{\underline{I}} + \underline{\underline{S}}$$

$$\underline{\underline{S}} = \mu \underline{\underline{A}}_1 + \alpha_1 \underline{\underline{A}}_2 + \alpha_2 \underline{\underline{A}}_1^2 + \beta_1 \underline{\underline{A}}_3 + \beta_2 \left(\underline{\underline{A}}_2 \underline{\underline{A}}_1 + \underline{\underline{A}}_1 \underline{\underline{A}}_2\right) + \beta_3 \left(\mathrm{tr}\, \underline{\underline{A}}_1^2\right) \underline{\underline{A}}_1$$

where $\underline{\underline{A}}_1 = \nabla \underline{v} + (\nabla \underline{v})^T \qquad \underline{\underline{A}}_n \equiv \frac{D \underline{\underline{A}}_{n-1}}{D} + \underline{\underline{A}}_{n-1} \nabla \underline{v} + (\nabla \underline{v})^T \underline{\underline{A}}_{n-1}$

The velocity dependence for laminar helical flow becomes $\underline{v} = (0, u_\theta(r)\; w(r))$, the stress which also is a function of r yields:

$$\underline{\underline{S}}_{rr} = (2\alpha_1 + \alpha_2)\left[\left(\frac{du_\theta}{dr} - \frac{u_\theta}{r}\right)^2 + \left(\frac{dw}{dr}\right)^2\right]$$

$$\underline{\underline{S}}_{r\theta} = \mu\left(\frac{du_\theta}{dr} - \frac{u_\theta}{r}\right) + 2\beta\left[\left(\frac{du_\theta}{dr} - \frac{u_\theta}{r}\right)^2 + \left(\frac{dw}{dr}\right)^2\right] \cdot \left(\frac{du_\theta}{dr} - \frac{u_\theta}{r}\right) \qquad \beta = \beta_1 +$$

$$\underline{\underline{S}}_{rz} = \mu \frac{dw}{dr} + 2\beta\left[\left(\frac{du_\theta}{dr} - \frac{u_\theta}{r}\right)^2 + \left(\frac{dw}{dr}\right)^2\right] \cdot \frac{dw}{dr}$$

$$\underline{\underline{S}}_{\theta\theta} = \alpha_2 \left(\frac{du_\theta}{dr} - \frac{u_\theta}{r}\right)^2$$

$$\underline{\underline{S}}_{\theta z} = \alpha_2 \left(\frac{du_\theta}{dr} - \frac{u_\theta}{r}\right) \frac{dw}{dr}$$

$$\underline{\underline{S}}_{zz} = \alpha_2 \left(\frac{dw}{dr}\right)^2$$

For the heat transfer in an isotropic, memory-less conductor,

$\underline{h} = -\kappa \nabla \Theta$ where the gradient components are $(\nabla\Theta)_r = \frac{\partial \Theta}{\partial r}$, $(\nabla\Theta)_\theta = \frac{1}{r}\frac{\partial \Theta}{\partial \theta}$, and $(\nabla\theta)_z = \frac{\partial \Theta}{\partial z}$.

Inserting the constitutive equations into the balance equations obtains:

$$\frac{1}{r}\frac{\partial}{\partial r}(ru_r)+\frac{1}{r}\frac{\partial u_\theta}{\partial \theta}+\frac{\partial w}{\partial z}=0 \qquad \text{(Total Mass)} \qquad (4)$$

$$\frac{d}{dr}(S_{rr})+\frac{S_{rr}-S_{\theta\theta}}{r}=\frac{\partial p}{\partial r}-\rho\frac{u_\theta^2}{r} \qquad \text{(r: Linear Momentum)} \qquad (5)$$

$$\frac{1}{r^2}\frac{d}{dr}\left[r^2 S_{r\theta}\right]=0 \qquad (\theta\text{: Linear Momentum}) \qquad (6)$$

$$\frac{1}{r}\frac{d}{dr}[rS_{rz}]=\frac{\partial p}{\partial z}\equiv -G \qquad \text{(z: Linear Momentum)} \qquad (7)$$

The temperature field is given by $\rho c_P \dfrac{D\Theta}{Dt}=\psi_v+\kappa\nabla\cdot(\nabla\Theta)=\psi_v+\kappa\nabla^2\Theta$

Here the viscous dissipation $\psi_r \equiv \mathrm{tr}\underline{\underline{S}}\cdot\underline{\underline{L}}=S_{rr}\dfrac{\partial u_r}{\partial r}+S_{\theta\theta}\left(\dfrac{1}{r}\dfrac{\partial u_\theta}{\partial r}\right)+S_{zz}\dfrac{\partial w}{\partial z}$

Then

$$\psi_v=(2\alpha_1+\alpha_2)\left[r\frac{d}{dr}\left(\frac{u_\theta}{r}\right)^2+\left(\frac{dw}{dr}\right)^2\right]\frac{du_r}{dr}+\alpha_2\left(r\frac{d}{dr}\left(\frac{u_\theta}{r}\right)\right)^2\left(\frac{1}{r}\frac{\partial u_\theta}{\partial r}\right)+\alpha_3\left(\frac{dw}{dr}\right)^2\frac{dw}{dz}$$

simplifies for helical flow to: $\alpha_2\left(r\dfrac{d}{dr}\left(\dfrac{u_\theta}{r}\right)\right)^2\left(\dfrac{1}{r}\dfrac{\partial u_\theta}{\partial r}\right)$.

Next, consider the isothermal problem first. The equations of motion (Eqns. 6 & 7) can be easily integrated to obtain: $S_{r\theta}=\dfrac{M}{r^2}$ (9)

$$S_{rz}=\frac{\partial p}{\partial z}\frac{r}{2}+\frac{N}{r} \qquad (10)$$

where M and N are constants of integration to be evaluated by the boundary conditions.

Using a perturbation approach for solution in terms of the small parameter β by letting,

$$\mathbf{u}_\theta = \mathbf{u}_{\theta 0} + \beta \mathbf{u}_{\theta 1} + \beta^2 \mathbf{u}_{\theta 2} + \ldots \quad \text{and} \quad \mathbf{w} = \mathbf{w}_0 + \beta \mathbf{w}_1 + \beta^2 \mathbf{w}_2 + .$$

The zero-th order system of equations derived from substituting the perturbed quantities into eqns. (9 & 10) can be solved subject to the following boundary conditions.

$$\mathbf{u}_\theta(\mathbf{r} = \mathbf{R}_1) = \Omega_1 \mathbf{R}_1 \qquad \mathbf{u}_\theta(\mathbf{r} = \mathbf{R}_2) = \Omega_2 \mathbf{R}_2$$

$$\mathbf{w}(\mathbf{r} = \mathbf{R}_2) = \overline{\mathbf{U}} \qquad \mathbf{w}(\mathbf{r} = \mathbf{R}_1) = 0$$

Since $\mu\left(\frac{\mathbf{du}_\theta}{\mathbf{dr}} - \frac{\mathbf{u}_\theta}{\mathbf{r}}\right) = \frac{\mathbf{M}}{\mathbf{r}^2}$, $\mu \mathbf{r}\frac{\mathbf{d}}{\mathbf{dr}}\left(\frac{\mathbf{u}_\theta}{\mathbf{r}}\right) = \frac{\mathbf{M}}{\mathbf{r}^2}$ and integrating

$\int_{\mathbf{R}_1}^{\mathbf{R}_2} \mathbf{d}\left(\frac{\mathbf{u}_\theta}{\mathbf{r}}\right) = \frac{\mathbf{M}}{\mu}\int_{\mathbf{R}_1}^{\mathbf{R}_2} \frac{1}{\mathbf{r}^3}\,\mathrm{dr}$ which becomes $\left.\frac{\mathbf{u}_\theta}{\mathbf{r}}\right|_{\mathbf{R}_1}^{\mathbf{R}_2} = \frac{\mathbf{u}_\theta(\mathbf{R}_2)}{\mathbf{R}_2} - \frac{\mathbf{u}_\theta(\mathbf{R}_1)}{\mathbf{R}_1} = \Omega_2 - \Omega_1 = \Delta\Omega$

$\Delta\Omega = \frac{\mathbf{M}}{-2\mu}\mathbf{r}^{-2}\Big|_{\mathbf{R}_1}^{\mathbf{R}_2} = \frac{-\mathbf{M}}{2\mu}\left(\frac{1}{\mathbf{R}_2^2} - \frac{1}{\mathbf{R}_1^2}\right)$. With known values for $\Delta\Omega$, $\mathbf{R}_1$, $\mathbf{R}_1$,

and a calibrating fluid for μ M can be found from $\mathbf{M} = 2\mu\Delta\Omega\frac{\mathbf{R}_1^2\mathbf{R}_2^2}{\mathbf{R}_2^2 - \mathbf{R}_1^2}$. Therefore,

$\int_{\mathbf{R}_1}^{\mathbf{r}} \mathbf{d}\left(\frac{\mathbf{u}_\theta}{\mathbf{r}}\right) = \frac{\mathbf{M}}{\mu}\int_{\mathbf{R}_1}^{\mathbf{r}} \frac{1}{\mathbf{r}^3}$ yields $\frac{\mathbf{u}_\theta(\mathbf{r})}{\mathbf{r}} = \Omega_1 + \frac{\mathbf{M}}{2\mu}\left(\frac{1}{\mathbf{R}_1^2} - \frac{1}{\mathbf{r}^2}\right)$ or re-writing

$$\mathbf{u}_\theta(\mathbf{r}) = \Omega_1 \mathbf{r} + \frac{\mathbf{M}}{2\mu}\left(\frac{\mathbf{r}}{\mathbf{R}_1^2} - \frac{1}{\mathbf{r}}\right).$$

The next order equation formed from the perturbation obtains:

$$\mu \mathbf{r}\frac{\mathbf{d}}{\mathbf{dr}}\left(\frac{\mathbf{u}_{\theta 1}}{\mathbf{r}}\right) + 2\beta \mathbf{r}\left[\mathbf{r}^2\left(\frac{\mathbf{d}}{\mathbf{dr}}\left(\frac{\mathbf{u}_{\theta 0}}{\mathbf{r}}\right)\right)^2 + \left(\frac{\mathbf{dw}_0}{\mathbf{dr}}\right)^2\right]\frac{\mathbf{d}}{\mathbf{dr}}\left(\frac{\mathbf{u}_{\theta 0}}{\mathbf{r}}\right) = \frac{\mathbf{M}}{\mathbf{r}^2}$$

This can be integrated as

$$\int_{R_1}^{r} d\left(\frac{u_{\theta 1}}{r}\right)=\frac{M}{\mu}\int_{R_1}^{r}\frac{dr}{r^3}+\frac{2\beta}{\mu}\int_{R_1}^{r}\left[r^2\left(\frac{d}{dr}\left(\frac{u_{\theta 0}}{r}\right)\right)^2+\left(\frac{dw_0}{dr}\right)^2\right]\frac{d}{dr}\left(\frac{u_{\theta 0}}{r}\right)dr \quad \text{so that}$$

$$u_\theta(r)=\Omega_1 r+\frac{M}{2\mu}\left(\frac{1}{R_1^2}-\frac{1}{r}\right)-I_1 r \quad \text{where}$$

$$I_1=\frac{2\beta}{\mu}\int_{R_1}^{r}\left[r^2\left(\frac{d}{dr}\left(\frac{u_{\theta 0}}{r}\right)\right)^2+\left(\frac{dw_0}{dr}\right)^2\right]\frac{d}{dr}\left(\frac{u_{\theta 0}}{r}\right)dr$$

The linear momentum eqn. 7 can also be solved with the perturbation series solution. In particular, re-writing and integrating with the constant N from integration as follows.

$$\frac{d}{dr}[rS_{rz}]=-Gr \quad \text{or} \quad \int d(rS_{rz})=-\int Gr\cdot dr=-\frac{G}{2}r^2+N \quad \text{so that}$$

$S_{rz}=-\frac{G}{2}r+\frac{N}{r}$. Then to zero-th order in β: $\mu\frac{dw_0}{dr}=-\frac{G}{2}r+\frac{N}{r}$.

Integration follows as $\int_{w(R_1)}^{w(R_2)}dw=-\frac{G}{2\mu}\int_{R_1}^{R_2}r\cdot dr+\frac{N}{\mu}\int_{R_1}^{R_2}\frac{dr}{r}$,

$w_0(R_2)-w_0(R_1)=\overline{U}-0=\frac{-G}{4\mu}(R_2^2-R_1^2)+\frac{N}{\mu}\ln\left(\frac{R_2}{R_1}\right)$. Solving for N,

$$N=\frac{\mu\left[\overline{U}+\frac{G}{4\mu}(R_2^2-R_1^2)\right]}{\ln\left(\frac{R_2}{R_1}\right)}$$

and $\int_{w(R_1)}^{w(r)}dw=-\frac{G}{2\mu}\int_{R_1}^{r}r\cdot dr+\frac{N}{\mu}\int_{R_1}^{r}\frac{dr}{r}$ obtains

$w_0(r)=\frac{-G}{4}(r^2-R_1^2)+\frac{N}{\mu}\ln\left(\frac{r}{R_1}\right)$. To first-order in β:

$$\frac{dw_1}{dr}+\frac{G}{2\mu}r-\frac{N}{\mu}\frac{1}{r}=\frac{-2\beta}{\mu}\left(r^2\left(\frac{d}{dr}\left(\frac{u_{\theta 0}}{r}\right)\right)^2+\left(\frac{dw_0}{dr}\right)^2\right)\frac{dw_0}{dr}$$

$$\int_{w(R_1)}^{w(r)} dw_1=-\frac{G}{2\mu}\int_{R_1}^{r} r\cdot dr+\frac{N}{\mu}\int_{R_1}^{r}\frac{dr}{r}-I_2 \text{ where}$$

$$I_2=\frac{2\beta}{\mu}\int\left(r^2\left(\frac{d}{dr}\left(\frac{u_{\theta 0}}{r}\right)\right)^2+\left(\frac{dw_0}{dr}\right)^2\right)\frac{dw_0}{dr}\cdot dr$$

Both I_1 and I_2 can easily be integrated with software such as Mathematica and calculations performed.

The volume flow rate Q can also be determined from integrating w over the cross-sectional area between the cylinders' radii. $Q=\int_0^{2\pi}\int_{R_1}^{R_2} w(r)\cdot r\cdot dr\cdot d\theta$

XV

Appendices

A. Equations of State

Gas Laws and Equations of State

Boyle's Law—temperature is constant.

$$P_1V_1 = P_2V_2$$

Charles' Law—pressure is constant.

$$V_1T_2 = V_2T_1 \text{ or } V_1/T_1 = V_2/T_2$$

Gay-Lussac's Law—volume is constant

$$P_1T_2 = P_2T_1$$

The Combined Gas Law:

$P_1V_1T_2 = P_2V_2T_1$ or $\frac{P_1V_1}{T_1} = \frac{P_2V_2}{T_2}$

Ideal Gas $PV = n\hat{R}T$ $PV=N\,k_BT$ $PV=MRT$

$$Pv=RT \text{ or } p=\rho RT$$

P = pressure
V = volume

n = number of moles = $\frac{M}{\hat{M}}$ = mass/molecular weight

V_m = V/n = *molar volume*
v=V/M
T = temperature (K)
$\hat{R}$ = universal gas constant = $N_A\ k_B$
R = *ideal gas constant* (8.314472 J/(mol·K))
k_B = Boltzmann constant (1.38054 10^{-23} J/°K/mole
N = number of molecules = n N_A
N_A= Avogadro's number (6.02252 10^{23} molecules/mole)

$\hat{M}$ = molecular mass = $\frac{M}{n}$ $n_i = \frac{M_i}{\hat{M}_i}$ $\quad n = \sum_{i=1}^{s} n_i$

R=$\frac{\hat{R}}{\hat{M}}$ = species gas constant

1 standard atmosphere = 101.3 kPascals = 760 mm Hg = 29.92 in Hg = 1.01325 bar = 14.7 lb/in^2

Ionization (e.g., Hydrogen)

$$n_H = n_{H^0} + n_{H^+} \qquad n_{H^+} \equiv n \qquad \alpha \equiv \frac{n_{H^+}}{n_H}$$

$$P = \left(n_{H^0} + n_e + n_{H^+}\right)kT = (1+\alpha)n_H kT$$

Radiation Pressure:

$P = \frac{1}{3}\alpha T^4$ Stefan's constant α = 7.565 10^{-16} JK^{-4}m^{-3}

Polytropic Equation (Thermally Perfect Gas):

$$\frac{p}{p_r} = \left(\frac{\rho}{\rho_r}\right)^{\gamma} \exp\left(\frac{s-s_r}{c_v}\right)$$, isentropic $s = s_r$ $\quad \frac{p}{p_r} = \left(\frac{\rho}{\rho_r}\right)^{\gamma}$ barotropic

Perfect Gas Mixture:

total mass $=\sum_{i=1}^{s} M_i$, total density $=\sum_{i=1}^{s} \rho_i = \frac{1}{V}\sum_{i=1}^{s} M_i = \frac{M}{V} = \rho$, mass fraction $x_i = \frac{M_i}{M}$

(Dalton's Law)

$$P = \sum_{i=1}^{s} p_i \quad p_i = \rho_i R_i T \qquad p_i = \frac{n_i}{n} P \qquad \frac{p_i}{p} = \frac{\rho_i}{\rho}\frac{R_i}{\hat{R}}$$

$$P = \sum_{i=1}^{s} \rho_i R_i T = \rho \hat{R} T \qquad R = \sum_{i=1}^{s} x_i R_i = \frac{\hat{R}}{\hat{M}} \qquad n_i = \frac{M_i}{\hat{M}_i} \qquad n = \sum_{i=1}^{s} n_i$$

$$\text{molecular mass of mixture } \frac{1}{\hat{M}} = \sum_{i=1}^{s} \frac{x_i}{\hat{M}_i}$$

Air with Variable Water Vapor

$$P = \rho R_{dryair} T_v \qquad T_v = \left(\frac{1+\frac{q}{e}}{1+q} \right) \qquad T_v \approx (1+0.608q)$$

q = mixing ratio (mass of water vapor per unit mass dry air)
e = ratio of the molecular weights of water vapor to dry air, respectively

Tait's Equation:

$$P = A\,[\,(\rho/\rho_o)^z - 1\,]$$

$z > 0$ is a material constant, and A is a "constant".
P. A. Thompson's *Compressible-Fluid Dynamics*

$$P = B\left[\left[\frac{\rho}{\rho_{sat}(T)}\right]^{\Gamma} - 1\right] + P_{sat}(T),\ B=3\ 10^8 \text{ bar},\ \Gamma=7$$

Schaffar, M., and H.J. Pfeifer. "Comparison of Two Computational Methods for High Velocity Cavitating Flows around Conical Projectiles." 2001.

$$\rho = \frac{\rho_0(T)}{1 - C(T)\ln\left(\frac{B(T)+P}{B(T)+P_0}\right)} \quad \rho_0 = \rho(P = P_0)$$ C, B functions of temperature

Vercher, S., and S. I. Anderson. "Measurement of Thermal Expansivity and Density as a Function of Pressure and Study of Isothermal Compressibility and Heat Capacity." Lungby, Denmark: Denmark Technical University.
Zhelezn, V. B. "Physical Model of the Structure of Harmonic Acoustic Waves In Liquid," Eighth Session of Russian Acoustical Society. (Moscow, August 25-29, 2003).

Berthelot's Equation: $P = \frac{RT}{V-b} - \frac{a}{TV^2}$

Modiïed Berthelot: $P = \frac{RT}{V}\left[1 + \frac{9PT_c}{128P_cT}\left(1 - \frac{6T_c^2}{T^2}\right)\right]$

Dieterici Equation: $P = \frac{RTe^{-a/(V_mRT)}}{V_m - b}$

Clausius Equation:

$$\left[P + \frac{a}{T(V_m + c)^2}\right](V_m - b) = RT \quad a = V_c - \frac{RT_c}{4P_c} \quad b = \frac{3RT_c}{8P_c} - V_c \quad c = \frac{27R^2T_c^3}{64P_c}$$

Van der Waals:

$$\left(P + \frac{a}{V_m^2}\right)(V_m - b) = RT$$ a, b empirical

$$a = \frac{27R^2T_C^2}{64P_c} \qquad b = \frac{RT_c}{8P_c}$$

T_c = critical temperature P_c = critical pressure

a is dependent upon the attractive forces between molecules while b is dependent upon repulsive forces.

Virial Equation of State:

$$\frac{PV_m}{RT} = 1 + \frac{B}{V_m} + \frac{C}{V_m^2} + \frac{D}{V_m^3} + \ldots$$

$$B = -V_c \qquad C = \frac{V_c^2}{3}$$

Redlich-Kwong Equation:

$$P = \frac{RT}{V_m - b} - \frac{a}{\sqrt{T}V_m\left(V_m + b\right)}$$

$$a = \frac{0.42748R^2T_c^{2.5}}{P_c}$$

$$b = \frac{0.08662RT_c}{P_c} \qquad \frac{P}{P_c} < 0.5 \cdot \frac{T}{T_c}$$

Redlich, O., and J. N. S. Kwong. *Chemistry Reviews* 44: (1949): 233.

Soave Equation:

$$P = \frac{RT}{V_m - b} - \frac{a\alpha}{V_m\left(V_m + b\right)}$$

$$a = \frac{0.42747R^2T_c^2}{P_c} \qquad b = \frac{0.08664RT_c}{P_c}$$

$$\alpha = \left(1 + \left(0.48508 + 1.55171\omega - 0.15613\omega^2\right)\left(1 - T_r^{0.5}\right)\right)^2$$

$$T_r = \frac{T}{T_c}$$

ω is the *acentric factor* for the species.

Peng-Robinson:

$$P = \frac{RT}{V_m - b} - \frac{a\alpha}{V_m^2 + 2bV_m - b^2}$$

$$a = \frac{0.45724R^2T_c^2}{P_c} \qquad b = \frac{0.07780RT_c}{P_c}$$

$$\alpha = \left(1 + \left(0.37464 + 1.54226\omega - 0.26992\omega^2\right)\left(1 - T_r^{0.5}\right)\right)^2$$

Peng, D. Y., and D. B. Robinson. *Industrial and Engineering Chemistry Fundamentals* 15 (1975): 59.

BWRS Benedict-Webb-Rubin

$$P = \rho RT + \left(B_0RT - A_0 - \frac{C_0}{T^2} + \frac{D_0}{T^3} - \frac{E_0}{T^4}\right)\rho^2 + \left(bRT - a - \frac{d}{T}\right)\rho^3 + \alpha\left(a + \frac{d}{T}\right)\rho^6 + \frac{c\rho^3}{T^2}\left(1 + \gamma\rho^2\right)\exp\left(-\gamma\rho^2\right)$$

ρ = the molar density

Starling, K.E. *Fluid Properties for Light Petroleum Systems.* Houston: Gulf Publishing Company, 1973.
Benedict, M., G. B. Webb, and L. C. Rubin. *Journal of Chemical Physics* 334 (1940).

Elliott, Suresh, Donohue:

$$\frac{PV_m}{RT} = Z = 1 + \frac{4\langle c\eta\rangle}{1 - 1.9\eta} - \frac{9.5\langle qY\eta\rangle}{1 + 1.7745\langle Y\eta\rangle}$$

c = shape factor

(repulsion term) $\eta = b \cdot \rho$, , b=component size parameter

$$q = 1 + 1.90476 \cdot (c - 1) \qquad Y = \exp\left(\frac{\epsilon}{kT}\right) - 1.0617$$

q = shape factor (attraction term)

Soule, A., C. Smith, X. Yang, and C. Lira. "Absorption Modeling with the ESD Equation of State." *Langmuir* 17 (2001): 2950-2957.

Beattie Bridgman:

$$P = \frac{RT\,(1-e)}{V^2}(V + B) - \frac{A}{V^2} \quad A = A_0\,(-a/V) \quad B = B_0\,(1 - b/V) \quad e = \frac{c}{VT^3}$$

Pitzer and Sterner Equation of State for Water:

$$\ln f = [\ln\rho + A^{res}/RT + P/\rho RT]_{P,T} + \ln(RT) - 1$$

f is fugacity; A^{res} is residual Helmholtz energy; ρ is 'molar' density (n/V); and P, T, R are pressure, temperature, and the universal gas constant, respectively.

Pitzer, K. S., and S. M. Sterner. "Equations of State Valid Continuously from Zero to Extreme Pressures for H2O and CO2." *Journal of Chemical Physics* 101(1994): 3111-3116.

Jones-Wilkens-Lee

$$P = Ae^{-R_1V} + Be^{-R_2V} + \frac{\omega c_v T}{V},$$

V = relative volume, ω = Gruneisen parameter, A, B, R_1, R_2 = constants

Lee, E. L., H. C. Horning, and J. W. King. "Adiabatic Expansion of High Explosive Detonation Products, UCRL-50422, Lawrence Livermore Laboratory, California, 1968, *Journal of Energetic Materials* 15, no. 4 (1997)289-311.

Mie-Gruneisen:

$$P = \frac{\Gamma c_v T}{V} + \Gamma \frac{f(V)}{V} g(V)$$

Lemons, D. S., and M. Lund. "Thermodynamics of High Temperature Mie-Gruneisen Solids." *Journal of Physics* 67 (December 1999): 12.

Davis, R. O., *American Journal of Physics* 40 (1972): 321-326.

Debeye Correction to Pressure:

$$P_{elec} = -\left(k\theta\right)/\left(24\pi D\right)) \qquad D^2 = \left(\varepsilon_0 k\theta\right)/\left(\sum_i n_i e^2 z_i^2\right)$$

summation over all ions and electrons

Bienkowski, G. K. "The Equations of State of an Ionized Gas." MIT Naval Supersonic Lab Report 398 (September 1959).

Equations of State Perturbations

Landau and Lifshitz: $P = p_0 + c_0^2 \rho^a - k_\theta\, p_\eta \theta_{,p} \nabla \bullet \underline{v}$

Polyakova, Soluyna, and Khokhlov Chemical Relaxation:

$$P = p_0 + c_0^2 \rho^a + 05\left(c^2\right)_{,p} \rho^{a2} + mc_0^2 \int e^{\Lambda t/\tau_1} \rho_{,t}^a dt_{past}$$

$$m = \frac{\left(c_\infty^2 - c_0^2\right)}{c_0^2} \qquad c_\infty^2 = p_{,p}\big|_z \qquad c_0^2 = p_{,p}\big|_\eta$$

BOSE EINSTEIN STATISTICS

Photons were modeled by Bose and atoms by Einstein with this distribution for indistinguishable particles whereby many particles can occupy the same lowest energy state at low temperature. The expected number in energy state i is given by:

$$\mathbf{n}_i = \frac{\mathbf{g}_i}{\mathbf{e}^{\frac{(\varepsilon_i - \mu)}{\mathbf{k}\theta}} - 1} \qquad \varepsilon_i - \mu \succ\succ \mathbf{k}\theta$$

$\mathbf{n}_i$ = number of particles in state i

ε_i = energy of state I, $\mathbf{g}_i$ = degeneracy i

μ = chemical potential

$\mathbf{k}$ = Boltzmann's constant

θ = absolute temperature

FERMI DIRAC DISTRIBUTION

For a single particle state the average number of Fermions is given by:

$$\mathbf{n}_i = \frac{1}{\mathbf{e}^{\frac{(\varepsilon_i - \mu)}{\mathbf{k}\theta}} + 1} \qquad 0 \prec \bar{\mathbf{n}}_i \prec 1$$

The average number of particles with energy state ε_i follows:

$$\mathbf{n}(\varepsilon_i) = \frac{\mathbf{g}_i}{\mathbf{e}^{\frac{(\varepsilon_i - \mu)}{\mathbf{k}\theta}} + 1}$$

Saha Equation (Weakly Ionized Gas)

The number of atoms in i-th state of ionization:

$$\frac{\mathbf{n}_{i+1} + \mathbf{n}_e}{\mathbf{n}_i} = \frac{2}{\Lambda^3} \frac{\mathbf{g}_{i+1}}{\mathbf{g}_i} \mathbf{e}^{-\frac{(\varepsilon_{i+1} - \varepsilon_i)}{\mathbf{k}_B\theta}} \qquad \Lambda \equiv \sqrt{\frac{\mathbf{h}^2}{2\pi \mathbf{m}_e \mathbf{k}_B \theta}}$$

$\mathbf{g}_i$ = degeneracy of states of i-th ions

ε_i = energy to remove I electrons from a neutral atom

$\mathbf{n}_e$ = electron number density

$\mathbf{h}$ = Planck's constant

$\mathbf{k}_B$ = Boltzmann's constant

$\mathbf{m}_e$ = mass of electron

θ : = temperature

M.N. Saha, "On A Physical Theory of Stellar Spectra," Proc. Royal Soc. London, Series A, *99*, 697 (May 1921).

Additional References

Tait, Peter Gunthrie. *Scientific Papers.* Cambridge: University Press, 1898.

Perrot, Pierre. *A to Z of Thermodynamics.* New York: Oxford University Press, 1998.

Elliott, J. R., and C. T. Lira. *Introductory Chemical Engineering Thermodynamics.* New Jersey: Prentice Hall, 1999.

Pauling, L. General Chemistry. New York: Dover, 1970.

Pitzer, Kenneth. *Thermodynamics (McGraw Hill Series in Advanced Chemistry).* New York: McGraw-Hill, 1995.

Atkins, P. W. *The Elements of Physical Chemistry.* Oxford: Oxford University Press, 1993.

Shah, K. K., and G. Thodos. *Industrial and Engineering Chemistry* 57, no. 3 (1965): 30.

Goodstein, D. L. *States of Matter.*New York: Dover, 1985.

Ross, M., and D. A. *Young.Annual Reviews of Physical Chemistry* 44 (1993): 61.

Span, R. *Multiparameter Equations of State.* Berlin: Springer, 2000.

Grossman, B. "Fundamental Concepts of Real Gas Dynamics." Lecture Notes, 2000.

Hess, S. L. *Introduction to Theoretical Meteorology.* New York: Holt, Rhinehart, Winston, 1959.

Chandrasekhar, S. "On Stars, Their Evolution, and Their Stability" Nobel Lecture, 1983.

B. Fractional Calculus

These brief notes introduce some of the derivations for generalizing integer calculus to fractional order such as a differential of one-half.

First, let's review the first differential as a limit of a differential quotient,

$$d^1 f(x) = \lim_{\Delta x \to 0} \frac{f(x+\Delta x) - f(x)}{\Delta x} .$$

This may be continued to generate the n-th derivative of a function (for natural number n)

$$d^n f(x) = \lim_{\Delta x \to 0} (\Delta x)^{-n} \sum_{m=0}^{n} (-1)^m \binom{n}{m} f\left(x + (n-m)\Delta x\right) \qquad \binom{n}{m} = \frac{n!}{m!(n-m)!}$$

or, $d^n f(x) = \lim_{\Delta x \to 0} (\Delta x)^{-n} \sum_{m=0}^{n} (-1)^m \binom{n}{m} f(x - m\Delta x)$ including n=0.

Noninteger values are obtained for the factorial by use of a gamma function.

The Grunwald-Letnikov derivative or differintegral is a generalization $\Delta x \to 0$, as *n* approaches infinity,

$$d^n f(x) = \lim_{\Delta x \to 0} (\Delta x)^{-n} \sum_{m=0}^{n} (-1)^m \binom{n}{m} f(x - m\Delta x) = \lim_{\Delta x \to 0} (\Delta x)^{-n} \sum_{m=0}^{n} (-1)^m \frac{n!}{m!(n-m+1)!} f(x - m\Delta x)$$

$$d^n f(x) = \lim_{\Delta x \to 0} (\Delta x)^{-n} \sum_{m=0}^{n \to \infty} (-1)^m \frac{n!}{m!(n-m+1)!} f(x - m\Delta x) = \lim_{\Delta x \to 0} (\Delta x)^{-n} \sum_{m=0}^{\frac{x-a}{\Delta x}} (-1)^m \frac{n!}{m!(n-m+1)!} f(x - m\Delta x)$$

$$d^\alpha f(x) = \lim_{\Delta x \to 0} (\Delta x)^{-n} \sum_{m=0}^{\frac{x-a}{\Delta x}} (-1)^m \frac{\Gamma(\alpha+1)}{m!\Gamma(\alpha-m+1)!} f(x - m\Delta x) = \lim_{n \to \infty} \left(\frac{n}{x-a}\right)^\alpha \sum_{m=0}^{n} (-1)^m \frac{\Gamma(\alpha+1)}{m!\Gamma(\alpha-m+1)!} f\left(x - m\left(\frac{x-a}{n}\right)\right)$$

where

$$\binom{-n}{m}=\frac{\prod_{L=n-m+1}^{-n} L}{m!}=\frac{\prod_{L=n}^{n+m-1}(-L)}{m!}=\frac{(-1)^m\prod_{L=n}^{n+m-1} L}{m!}=\frac{(-1)^m\Gamma(n+m-1)!}{m!(n-1)!}=\frac{(-1)^m\Gamma(n+m)!}{m!\Gamma(n)}$$

Alternatively, the Riemann-Louiville formula is derived by the method of iterated integrals.

The first integral of the function f is denoted by a (-1) differential.

$$d^{-1} f(x)=\int_0^x f(t)dt$$

Also,

$$d^{-2} f(x)=\int_0^x\int_0^{t_2} f(t_1)dt_1dt_2=\int_0^x\int_{t_1}^x f(t_1)dt_2dt_1=\int_0^x f(t_1)\int_{t_1}^x dt_2dt_1=\int_0^x f(t)(x-t)dt\ .$$

Repeating this process for iterated integrals $d^{-n} f(x)=\frac{1}{(n-1)!}\int_0^x f(t)(x-t)^{n-1}dt$

and generalizing to noninteger values, $d^{\alpha} f(x)=\frac{1}{\Gamma(-\alpha)}\int_0^x \frac{f(t)}{(x-t)^{\alpha+1}}dt$.

Several properties and selected examples are shown below,

$$d^{k+n} f(x)=d^k d^n f(x)$$
$$d^k(af(x)+bg(x))=ad^k f(x)+bd^k g(x)$$

$$d^k e^{ax}=a^k e^{ax}$$

$$d^k\cos(x)+id^k\sin(x)=d^k e^{ix}=i^k e^{ix}=e^{\frac{k\pi i}{2}}e^{ix}=e^{i(x+k\pi/2)}$$
$$=\cos\left(x+\frac{k\pi}{2}\right)+i\sin\left(x+\frac{k\pi}{2}\right)$$

$$d^k x^a = \frac{\Gamma(a+1)}{\Gamma(a-k+1)} x^{a-k}$$

For a power series function, $f(x) = \sum_{n=-\infty}^{\infty} a_n x^n$

$$d^k f(x) = \sum_{n=-\infty}^{\infty} d^k a_n x^n = \sum_{n=-\infty}^{\infty} a_n \frac{\Gamma(n+1)}{\Gamma(n-k+1)} x^{n-k}$$

then the familiar integer differential Taylor series can be generalized,

$$f(x+a) = \sum_{n=0}^{\infty} \frac{d^n f(a)}{n!} x^n \quad \rightarrow \quad d^k f(x+a) = \sum_{n=0}^{\infty} \frac{d^n f(a)}{\Gamma(n-k+1)} x^{n-k}$$

C. Compressible Bubble Dynamics Motion

Define the enthalpy $\mathbf{H} = \mathbf{e} + \mathbf{pv}$ $\quad \mathbf{v} = \frac{1}{\rho}$ where $\mathbf{e}$ is the internal energy and $\mathbf{v}$ the specific volume or reciprocal of density ρ. The differential $\mathbf{dH} = \mathbf{de} + \mathbf{pdv} + \mathbf{vdp}$.

Neglecting viscous and heat dissipation contributions to entropy, $\mathbf{d\eta} = \mathbf{de} + \mathbf{pdv} = 0$ obtains, $\rho\frac{\mathbf{dH}}{\mathbf{dt}} = \frac{\mathbf{dp}}{\mathbf{dt}}$, or $\mathbf{dH} = \frac{1}{\rho}\frac{\mathbf{dp}}{\mathbf{dt}}\mathbf{dt}$. Then integrating $\mathbf{H} = \int_{\mathbf{R}}^{\infty}\mathbf{dH} = \int_{\mathbf{R}}^{\infty}\frac{1}{\rho}\frac{\mathbf{dp}}{\mathbf{dt}}\mathbf{dt}$

For a spherically symmetric bubble,

$$\mathbf{p(R,t)}; \qquad \frac{\partial}{\partial \mathbf{R}} = \frac{1}{\mathbf{c}}\frac{\partial}{\partial \mathbf{t}} \qquad \frac{\mathbf{dp}}{\mathbf{dt}} = \frac{\partial \mathbf{p}}{\partial \mathbf{R}}\frac{\partial \mathbf{R}}{\partial \mathbf{t}} + \frac{\partial \mathbf{p}}{\partial \mathbf{t}}$$

The bubble dynamics are simply $\mathbf{R\ddot{R}} + \frac{3}{2}\mathbf{\dot{R}}^2 = -\mathbf{H}$ relating the kinetic energy and work $\frac{1}{2}\rho\int_{\mathbf{R}}^{\infty}\mathbf{\dot{r}}^2\left(4\pi\mathbf{r}^2\right)\mathbf{dr} = \int_{\mathbf{R}_0}^{\mathbf{R}}(\mathbf{p}_{\mathbf{L}} - \mathbf{p}_{\infty})4\pi\mathbf{R}^2\mathbf{dR}$ and noting from continuity that $\frac{\mathbf{\dot{r}}}{\mathbf{\dot{R}}} = \frac{\mathbf{R}^2}{\mathbf{r}^2}$ or $\mathbf{\dot{r}} = \frac{\mathbf{R}^2\mathbf{\dot{R}}}{\mathbf{r}^2}$,

$$\frac{\partial}{\partial \mathbf{R}}\int_{\mathbf{R}_0}^{\mathbf{R}}(\mathbf{p}_{\mathbf{L}} - \mathbf{p}_{\infty})4\pi\mathbf{R}^2\mathbf{dR} = \frac{\partial}{\partial \mathbf{R}}\left(2\pi\mathbf{R}^3\mathbf{\dot{R}}^2\rho\right) .$$

A viscous contribution on the right-hand side derives from the balance of linear momentum as follows, $-\int_{\mathbf{R}_0}^{\infty}\nabla\cdot\underline{\underline{\mathbf{S}}}\cdot\mathbf{dr}$ integrating the stress tensor of the liquid.

The special case of constant density and a perfect inviscid fluid with no surface tension,

$$\mathbf{R\ddot{R}} + \frac{3}{2}\mathbf{\dot{R}}^2 = -\frac{\Delta \mathbf{p}}{\rho_0} \equiv \mathbf{cst} = \frac{\mathbf{p}_L - \mathbf{p}_\infty}{\rho_0}$$. Let

$$\mathbf{P} = \mathbf{\dot{R}}, \qquad \mathbf{\dot{P}} = \mathbf{\ddot{R}}, \qquad \mathbf{\dot{P}} = \frac{\partial \mathbf{P}}{\partial \mathbf{R}}\frac{\partial \mathbf{R}}{\partial \mathbf{t}} = \mathbf{P'\dot{R}}$$

Then $\mathbf{R\dot{P}} + \frac{3}{2}\mathbf{P}^2 = \alpha\,(\mathrm{cst})$, $\frac{\mathbf{PdP}}{\alpha - \frac{3}{2}\mathbf{P}^2} = \frac{\mathbf{dR}}{\mathbf{R}}$, and $\ell\mathbf{nRC} = -\frac{1}{3}\ell\mathbf{n}\left(\alpha - \frac{3}{2}\mathbf{P}^2\right)$ or

$\mathbf{RC} = \left(\alpha - \frac{3}{2}\mathbf{P}^2\right)^{-\frac{1}{3}}$ with initial conditions to solve for the integration constant C,

$\mathbf{\dot{R}} = 0$ at $t = 0$, so $\mathbf{C} = \frac{1}{\mathbf{R}_0\sqrt[3]{\alpha}}$, $\mathbf{P} = \mathbf{\dot{R}} = -\sqrt{\frac{2}{3}\frac{\Delta \mathbf{p}}{\rho_0}\left(\frac{\mathbf{R}_0^3}{\mathbf{R}^3} - 1\right)}$. The time to collapse is

$$\mathbf{t} = \mathbf{R}_0\sqrt{\frac{3}{2}\frac{\rho_0}{\mathbf{p}_\infty}}\int_\beta^1 \frac{\beta^{3/2}\mathbf{d}\beta}{\sqrt{1-\beta^3}}$$, $\beta = 0$ is complete collapse, $\beta \equiv \frac{\mathbf{R}}{\mathbf{R}_0}$; $\mathbf{R} = 0$.

For a spherical bubble in a Newtonian fluid with surface tension,

$$\rho\left(\mathbf{R\ddot{R}} + \frac{3}{2}\mathbf{\dot{R}}^2\right) = \Delta \mathbf{p} - 4\mu\frac{\mathbf{\dot{R}}}{\mathbf{R}} - 2\frac{\sigma}{\mathbf{R}}$$

Balancing normal forces at the interface of the bubble, and modifying the bubble equation of motion for a viscous liquid obtains,

$$-\int_{\mathbf{R}_0}^\infty \nabla \cdot \underline{\underline{\mathbf{S}}} \cdot \mathbf{dr} = -\int_{\mathbf{R}_0}^\infty \left(\frac{1}{\mathbf{r}^2}\frac{\partial}{\partial \mathbf{r}}(\mathbf{r}^2\mathbf{S}_{rr}) + \frac{1}{\mathbf{r}\sin\theta}\frac{\partial}{\partial\theta}(\mathbf{S}_{\theta r}\sin\theta) + \frac{1}{\mathbf{e}\sin\theta}\frac{\partial \mathbf{S}_{\varphi r}}{\partial\varphi} - \frac{\mathbf{S}_{\theta\theta} - \mathbf{S}_{\varphi\varphi}}{\mathbf{r}}\right)\cdot \mathbf{dr}$$

$$= -\int_{\mathbf{R}_0}^\infty \left(\frac{1}{\mathbf{r}^2}\frac{\partial}{\partial \mathbf{r}}(\mathbf{r}^2\mathbf{S}_{rr}) - \frac{\mathbf{S}_{\theta\theta}}{\mathbf{r}}\right)\cdot \mathbf{dr} \quad -2\int_{\mathbf{R}_0}^\infty \frac{\mathbf{S}_{rr} - \mathbf{S}_{\theta\theta}}{\mathbf{r}}\mathbf{dr}$$

$$\mathbf{R\ddot{R}} + \frac{3}{2}\mathbf{\dot{R}}^2 = \mathbf{H} - \frac{2\sigma}{\rho \mathbf{r}} - 2\int_{\mathbf{R}_0}^\infty \frac{1}{\rho}\frac{\mathbf{S}_{rr} - \mathbf{S}_{\theta\theta}}{\mathbf{r}}\mathbf{dr} - \frac{\nabla\sigma}{\rho}$$

Define a potential by $\nabla\phi = \mathbf{v}_r$, then $\frac{\partial\varphi}{\partial t}+\frac{1}{2}\left(\frac{\partial\varphi}{\partial \mathbf{r}}\right)^2 = -\int\frac{\mathbf{dp}}{\rho}+\int\frac{1}{\rho}\underline{\mathbf{n}}\cdot\nabla\cdot\mathbf{S}\cdot\underline{\mathbf{n}}\mathbf{dr}$,

$$\phi = \frac{\dot{\mathbf{R}}\mathbf{R}^2}{\mathbf{r}} + \phi_\infty$$

For constant density, $\int\frac{1}{\rho}\underline{\mathbf{n}}\cdot\nabla\cdot\mathbf{S}\cdot\underline{\mathbf{n}}\mathbf{dr} = \frac{1}{\rho}\left[\frac{2\sigma}{\mathbf{R}} - \nabla\sigma\right]$ $\quad \nabla\cdot\underline{\mathbf{n}} = \kappa = \frac{1}{\mathbf{R}_1}+\frac{1}{\mathbf{R}_2} \to \frac{2}{\mathbf{R}}$

Let the extra stress

$$\mathbf{S}_{\mathbf{rr}} = \mathbf{E}_0\frac{\partial\mathbf{u}_\mathbf{r}}{\partial\mathbf{r}} + \mathbf{E}_1\frac{\partial}{\partial\mathbf{r}}\frac{\partial^{\alpha+1}\mathbf{u}_\mathbf{r}}{\partial\mathbf{t}^{\alpha+1}} = \mathbf{E}_0\frac{\partial\mathbf{u}_\mathbf{r}}{\partial\mathbf{r}} + \mathbf{E}_1\frac{\partial}{\partial\mathbf{r}}\frac{\partial^{\alpha}\mathbf{v}_\mathbf{r}}{\partial\mathbf{t}^{\alpha}}, \qquad \mathbf{v}_\mathbf{r} = \frac{\partial\mathbf{u}_\mathbf{r}}{\partial\mathbf{t}}$$

for fractional differ-integral order α denote the extra-stress. Then $\mathbf{S}_{\mathbf{rr}} = \mathbf{E}_0\varphi + \mathbf{E}_1\frac{\partial^\alpha\varphi}{\partial\mathbf{t}^\alpha}$ in terms of the potential function and

$$\frac{\partial\varphi}{\partial t}+\frac{1}{2}\left(\frac{\partial\varphi}{\partial \mathbf{r}}\right)^2 = -\frac{1}{\rho}\int\mathbf{dp} + \mathbf{E}_0\varphi + \mathbf{E}_1\frac{\partial^\alpha\varphi}{\partial\mathbf{t}^\alpha} - \frac{2\sigma}{\mathbf{R}} - \nabla\sigma \quad .$$

Assuming and neglecting surface tension gradients, let $\mathbf{E}_0 = 0$, with a transformation of coordinates (presently c=1)

$\mathbf{z} \equiv \mathbf{r} - \mathbf{ct}$, $\quad \frac{\partial}{\partial\mathbf{r}} = \frac{\partial}{\partial\mathbf{z}}$, $\quad \frac{\partial}{\partial\mathbf{t}} = -\mathbf{c}\frac{\partial}{\partial\mathbf{z}}$ enables one to re-write the partial differential equation of surface motions as follows.

$$-\mathbf{c}\frac{\partial\phi}{\partial\mathbf{z}}+\frac{1}{2}\left(\frac{\partial\phi}{\partial\mathbf{z}}\right)^2 = -\mathbf{P} + (-\mathbf{c})\mathbf{E}_1\frac{\partial^{\alpha-1}\partial\phi}{\partial\mathbf{t}^{\alpha-1}\partial\mathbf{z}} \text{ where } \mathbf{P} = \frac{1}{\rho}\Delta\mathbf{p} + \frac{2\sigma}{\mathbf{R}}, \qquad \int\mathbf{dp} = \Delta\mathbf{p} \ .$$

Now define $\mathbf{V} = \frac{\partial\phi}{\partial\mathbf{z}}$ so that $-\mathbf{cV} + \frac{1}{2}(\mathbf{V})^2 + \mathbf{P} = -(-\mathbf{c})\mathbf{E}_1\frac{\partial^{\alpha-1}\mathbf{V}}{\partial\mathbf{z}^{\alpha-1}} - \nabla\sigma$

Neglecting gradients of surface tension yields,

$$(\mathbf{V})^2 - 2\mathbf{cV} + 2\mathbf{P} = (-2\mathbf{c})\mathbf{E}_1\frac{\partial^{\alpha-1}\mathbf{V}}{\partial\mathbf{z}^{\alpha-1}}$$

Completing the square $(\mathbf{V})^2 - 2\mathbf{cV} + 2\mathbf{P} = (\mathbf{V}-\mathbf{c})^2 + \mathbf{c}^2 + 2\mathbf{P}$ or

$$= (\mathbf{V}-\mathbf{c})^2 + \Lambda \text{ with } \beta \equiv (-2\mathbf{c})\mathbf{E}_1 \ ,$$

$\Lambda = \mathbf{c}^2 + 2\mathbf{P}$, and obtains $(\mathbf{V}-\mathbf{c})^2 + \Lambda = \beta \dfrac{\partial^{\alpha-1}\mathbf{V}}{\partial \mathbf{z}^{\alpha-1}}$

The fractional order differential equation for the bubble becomes

$$\left(\frac{\mathbf{V}}{\sqrt{\beta}} - \frac{\mathbf{c}}{\sqrt{\beta}}\right)^2 + \frac{\Lambda}{\beta} = \frac{\partial^{\alpha-1}\mathbf{V}}{\partial \mathbf{z}^{\alpha-1}} \ .$$

Further investigations would involve both heat and mass transfer fluxes to be balanced at the interface accommodating transfers between the exterior and interior of the bubble where fractional calculus may also be used in the formulation.

The familiar form of the above equation is $\alpha = 2$ so that

$$\frac{\partial \mathbf{V}^{\alpha-2}\partial \mathbf{V}}{\left(\frac{\mathbf{V}}{\sqrt{\beta}} - \frac{\mathbf{c}}{\sqrt{\beta}}\right)^2 + \frac{\Lambda}{\beta}} = \partial^{\alpha-2}\partial \mathbf{z} \quad \text{obtains}$$

$\displaystyle\int \frac{\mathbf{dV}}{\left(\frac{\mathbf{V}}{\sqrt{\beta}} - \frac{\mathbf{c}}{\sqrt{\beta}}\right)^2 + \frac{\Lambda}{\beta}} = \int \mathbf{dz}$ which is integrated by absolute integer calculus with substitution and a trigonometry.

$\displaystyle\int \frac{\mathbf{dV}}{\left(\frac{\mathbf{V}}{\sqrt{\beta}} - \frac{\mathbf{c}}{\sqrt{\beta}}\right)^2 + \frac{\Lambda}{\beta}} = \int \mathbf{dz}$ First substitute $\Omega = \left(\frac{\mathbf{V}}{\sqrt{\beta}} - \frac{\mathbf{c}}{\sqrt{\beta}}\right)$;

$\mathbf{k} = \dfrac{\Lambda}{\beta} \qquad \mathbf{dV} = \sqrt{\beta}\mathbf{d}\Omega \qquad \sqrt{\beta}\displaystyle\int \frac{\mathbf{d}\Omega}{\Omega^2 + \mathbf{k}} = \int \mathbf{dz}$ Furthermore,

$$\frac{\sqrt{\beta}}{\mathbf{k}}\int\frac{\mathbf{d}\Omega}{\left(\frac{\Omega}{\sqrt{\mathbf{k}}}\right)^2+1}=\int\mathbf{dz}\quad \text{Let}\quad Z=\frac{\Omega}{\sqrt{\mathbf{k}}},\qquad \sqrt{\mathbf{k}}\mathbf{dZ}=\mathbf{d}\Omega$$

so that the integral can be re-written as $\sqrt{\frac{\beta}{\mathbf{k}}}\int\frac{\mathbf{dZ}}{\mathbf{Z}^2+1}=\int\mathbf{dz}$.

For $\mathbf{Z}=\tan\Theta$, $\mathbf{dZ}=\sec^2\Theta\mathbf{d}\Theta$, $\sqrt{\frac{\beta}{\mathbf{k}}}\int\frac{\sec^2\mathbf{d}\Theta}{\tan\Theta^2+1}=\sqrt{\frac{\beta}{\mathbf{k}}}\int\mathbf{d}\Theta$

$$\mathbf{z}=\sqrt{\frac{\beta}{\mathbf{k}}}\Theta,\qquad \Theta=\tan^{-1}\mathbf{Z}=\sqrt{\frac{\mathbf{k}}{\beta}}\mathbf{z}\quad \mathbf{Z}=\tan\left(\sqrt{\frac{\mathbf{k}}{\beta}}\mathbf{z}\right)=\frac{\Omega}{\sqrt{\mathbf{k}}}$$

$$\Omega=\sqrt{\frac{\Lambda}{\beta}}\tan\left(\frac{1}{\beta}\sqrt{\Lambda}\mathbf{z}\right)\ \text{and}\ \mathbf{V}=\mathbf{c}+\sqrt{\Lambda}\tan\left(\frac{1}{\beta}\sqrt{\Lambda}\mathbf{z}\right)$$

$$\frac{\mathbf{d}\varphi}{\mathbf{dz}}=\mathbf{c}+\sqrt{\Lambda}\,\mathbf{tan}\left(\frac{1}{\beta}\sqrt{\Lambda}\mathbf{z}\right),\quad \int\mathbf{d}\varphi=\int\mathbf{cdz}+\sqrt{\Lambda}\int\mathbf{tan}\left(\frac{1}{\beta}\sqrt{\Lambda}\mathbf{z}\right)\mathbf{dz}$$

Finally, the potential

$$\phi=\mathbf{z}+\frac{\Lambda}{\beta}\mathrm{h}\left|\sec\frac{\sqrt{\Lambda}}{\beta}\mathbf{z}\right|\quad\text{or}\quad \phi=(\mathbf{r}-\mathbf{t})+\frac{\Lambda}{\beta}\mathrm{h}\left|\sec\frac{\sqrt{\Lambda}}{\beta}(\mathbf{r}-\mathbf{t})\right|+\text{constant}$$

$$\Lambda=1+\left(\frac{2}{\rho}\Delta\mathbf{p}+\frac{4\sigma}{\mathbf{R}}\right)$$

D. On The Wire Coating Equation

In a recent publication, the following ordinary differential equation was derived and numerically solved. The following represents an alternative analytical solution approach using substitution, integration, and algebra.

Consider $\frac{d}{dr}\left(r\frac{du}{dr}\right)-\alpha\frac{d}{dr}\left[r\left(\frac{du}{dr}\right)^3\right]=0 \qquad -\alpha\equiv\frac{2(\beta_2+\beta_3)}{\mu}$

Integrating, $r\frac{du}{dr}-\alpha r\left(\frac{du}{dr}\right)^3=\kappa$ **(constant)** which may be rewritten as

$\frac{du}{dr}-\alpha\left(\frac{du}{dr}\right)^3=\frac{\kappa}{r}$ or letting $V=\frac{du}{dr}$, $V-\alpha V^3=\frac{\kappa}{r}=0$ since $\frac{du}{dr}=V=0$

at $r=R_w$ because $u=U_\infty$ a constant. For $R_w\le r<R_d$ in order not to

divide by zero $\frac{du}{dr}-\alpha\left(\frac{du}{dr}\right)^3=0$ becomes $1-\alpha\left(\frac{du}{dr}\right)^2=0$ which can

be solved $\left(\frac{du}{dr}\right)=\pm\sqrt{\frac{1}{\alpha}}=\pm\alpha^{-0.5}$

$\int_{u(R_w)}^{u(r)} du=u(r)-U_W=\pm\alpha^{-0.5}\int_{R_w}^{r} dr=\alpha^{-0.5}(r-R_w)$ The condition

that $u\ge 0$ can be used to discard the negative solution.

This positive sloping linear variation of velocity with radial position r is exhibited in the calculations in Fig. 4 This can be used to calculate the radius of the coated wire and force required to pull the wire which follows.

$$R_C=\left[R_W^2+\frac{2}{U_W}\left(\frac{r^2}{2}\left[U_W-\frac{R_W}{\sqrt{\alpha}}\right]+\frac{r^3}{3\sqrt{\alpha}}\right)\right]^{0.5}-\left[R_W^2+\frac{2}{U_W}\left(\frac{R_W^2}{2}\left[U_W-\frac{R_W}{\sqrt{\alpha}}\right]+\frac{R_W^3}{3\sqrt{\alpha}}\right)\right]^{0.5}$$

$$\mathbf{T}_{rz} = \frac{\mu}{\sqrt{\alpha}}\left(1+\frac{2\beta}{\alpha}\right) \text{ and } \mathbf{F}_W = 2\pi\frac{\mathbf{R}_W\mathbf{L}\mu}{\sqrt{\alpha}}\left(1+\frac{2\beta}{\alpha}\right) \text{ where } \beta = \beta_2 + \beta_3$$

Reference

"Wire Coating By Withdrawal from A Bath of Fourth Order by A.M. Siddiqui, M. Sajid, and T. Hayat, *Physics Letter A* 372, pgs-2665-2670 (2007).

E: Statistical Hydrodynamics:

Fractional calculus along with probabilistic "particle jumps of length and direction" or continuous time random walks as compound Poisson models appear most useful for statistical hydrodynamic problems. In hydrogeology, fractional order multivariable derivatives for dispersion terms have been introduced in the equation for predicting concentrations from a source release in a porous medium.

$$\frac{\partial \mathbf{C}(\underline{\mathbf{x}},\mathbf{t})}{\partial \mathbf{t}} + \underline{\mathbf{v}} \cdot \nabla \mathbf{C} = \mathbf{D}\nabla^{\alpha}\mathbf{C}$$

The concentration patterns are viewed as convolutions of concentrations

with power laws with range. $\frac{\mathbf{d}^{n}\mathbf{C}(\mathbf{x},\mathbf{t})}{\mathbf{dt}^{n}} = \frac{1}{\Gamma(-\alpha)}\int \mathbf{r}^{-\alpha-1}\mathbf{C}(\mathbf{x}-\mathbf{r})\mathbf{dr}$

Further important developments in scaling and transforming to account for anisotropic patterns were also addressed.

References:

Meerschaert, M. M., D.A. Benson, Schumer, R, Baeumer, B., *Phys. Rev.* E, 65, 1103 (2002).

Benson, D. A., S.W. Wheatcraft, and M. M. Meerschaert, *Water Resources Research, 36,* 1403 (2000).

Schumer, R., D. A. Benson, Meerschaert, M. M., Baeumer, B., H. P. Scheffler, *Water Resources Research*, 39, 1, 1022 (2003).

Gorenflo, R. and F. Mainiardi, *Fractals and Fractional Calculus in Continuum Mechanics*, 223-276, Springer (1997).

Gorenflo, R. and F. Mainiardi, *Frac. Calc. Appl. Anal.*, 4, 153-192 (2001).

About The Author

Timothy Scott Margulies was educated at The Johns Hopkins University in Baltimore, Maryland and earned his Ph.D in Applied Mathematics and Fluid Mechanics.

About The Book

The book presents research and lecture notes from both university teaching and work experiences.

www.ingramcontent.com/pod-product-compliance
Lightning Source LLC
Chambersburg PA
CBHW032026170526
45157CB00002B/873
* 9 7 8 1 4 3 6 3 9 9 0 1 2 *